Clinical Performance of Skeletal Prostheses

Clinical Performance of Skeletal Prostheses

Edited by

Larry L. Hench and June Wilson

SPRINGER-SCIENCE+BUSINESS MEDIA, B.V

First edition 1996

Originally published by Chapman & Hall in 1996

ISBN 978-0-412-72110-6   ISBN 978-94-011-0541-5 (eBook)
DOI 10.1007/978-94-011-0541-5

A catalogue record for this book is available from the British Library.

Editorial assistants: *Alice Holt*, *Waverly Munn* and *Jon West*

♾Printed on acid-free text paper, manufactured in accordance with ANSI/NISO Z39.48-1992 (Permanence of Paper).

Contents

Contributors

Sameer Bhatia, M.D.
Material Science and Engineering
University of Florida
Gainesville, FL 32611

John Bianchi, M.S.
University of Florida Tissue Bank
One Progress Blvd, Box 31
Alachua, FL 32615

Joan Chan, B.S.
University of Florida Tissue Bank
One Progress Blvd, Box 31
Alachua, FL 32615

Arthur E. Clark PhD, D.D.S.
Department of Prosthodontics
College of Dentistry
University of Florida
Gainesville, FL 32611

Charles F. De Freest, D.D.S.
Dental Biomaterials
College of Dentistry
University of Florida
Gainesville, FL 32610

John Delvaux, M.S.
United Technologies
Pratt & Whitney
P.O. Box 109600
West Palm Beach, FL 33140

Shannon Eggers, B.S.
314 N.W. 1st St, Apt A
Gainesville, FL 32601

Laura Elsberg, B.S.
Material Science and Engineering
University of Florida
Gainesville, FL 32611

Shawn H. Gallagher, B.S.
3425 S.W. 2nd Ave, Apt 158
Gainesville, FL 32607

Neil Graf, B.S.
314 N.W. 1st St, Apt A
Gainesville, FL 32601

Jason A. Griggs, B.S.
3425 S.W. 2nd Ave, Apt 158
Gainesville, FL 32607

Jamie M. Grooms, B.S.
University of Florida Tissue Bank
One Progress Blvd, Box 31
Alachua, FL 32615

Larry L. Hench, PhD
Material Science and Engineering
University of Florida
Gainesville, FL 32611

Thomas J. Hill, B.S.
3425 S.W. 2nd Ave, Apt 158
Gainesville, FL 32607

Daniel F. Justin, M.S.
Matthews Orthopedic Clinic
1315 S. Orange Ave
Orlando, FL 32806

Penelope E. Kao, B.S.
314 N.W. 1st St, Apt A
Gainesville, FL 32601

Adam Leckey, B.S.
3746 S.W. 3rd Place
Gainesville, FL 32607

Bernd Liesenfield, M.S.
314 N.W. 1st St, Apt A
Gainesville, FL 32601

Keith D. Lobel, B.S.
Material Science and Engineering
University of Florida
Gainesville, FL 32611

Julie A. Miller, M.S.
Material Science and Engineering
University of Florida
Gainesville, FL 32611

Mark Moore, B.S.
Material Science and Engineering
University of Florida
Gainesville, FL 32611

Rodrigo L. Orefice, M.S.
Material Science and Engineering
University of Florida
Gainesville, FL 32611

Sheila Rao, B.S.
3746 S.W. 3rd Place
Gainesville, FL 32607

Daniel A. Savett, D.M.D
Dental Biomaterials
College of Dentistry
University of Florida
Gainesville, FL 32610

Teresa Schimmel, B.S.
3425 S.W. 2nd Ave, Apt 158
Gainesville, FL 32607

Greg Snyder, B.S.
United Technologies
Pratt & Whitney
P.O. Box 109600
West Palm Beach, FL 33140

Harold R. Stanley, D.D.S.
Two Sea Oaks Terr.
Ormond Beach, FL 32074

James D. Talton, B.S.
Material Science and Engineering
University of Florida
Gainesville, FL 32611

June Wilson, PhD
Bioglass Research Center
College of Dentistry
University of Florida
Gainesville, FL 32610

1

Introduction

Larry L. Hench
June Wilson

OBJECTIVE

Millions of people presently enjoy an improved quality of life due to prostheses which repair, augment or replace parts of their skeletal system: bones, joints, teeth, etc. However, all replacement parts have a finite probability of survival. The goal of this book is to compare the survivability data for various skeletal prosthesis systems. All data derive from previously published clinical studies. Where possible statistical comparisons are made and the reasons for failure are discussed.

THE NEED FOR SKELETAL PROSTHESES

We are an aging population with more than 100 million people in the U.S. and Europe over the age of 50 years. An unfortunate consequence of aging is a progressive deterioration of the quality of skeletal tissues. From the age of 30 years there is a decrease in bone mass for both men and women (Fig. 1.1). However, for women it is much greater and between 40 and 60 years of age the rate of deterioration of long bones and vertebrae of women is especially severe due to hormonal changes. By the age of 70 most women will have lost from 35 to 60% of their bone mass. The loss of volume of cancellous or trabecular bone leads to a large decrease in mechanical compressive strength (Fig. 1.2). The clinical consequence is an increasing incidence of vertebral collapse. Cortical bone decreases in tensile strength with age (Fig. 1.2), which, combined with loss of cancellous bone, makes the neck of the femur particularly susceptible to fracture. The incidence of broken long bones and joints (especially the hip) thus increases with age.

The effects of aging, together with an increased incidence of other diseases of bones, joints and teeth, have led to an enormous growth in the

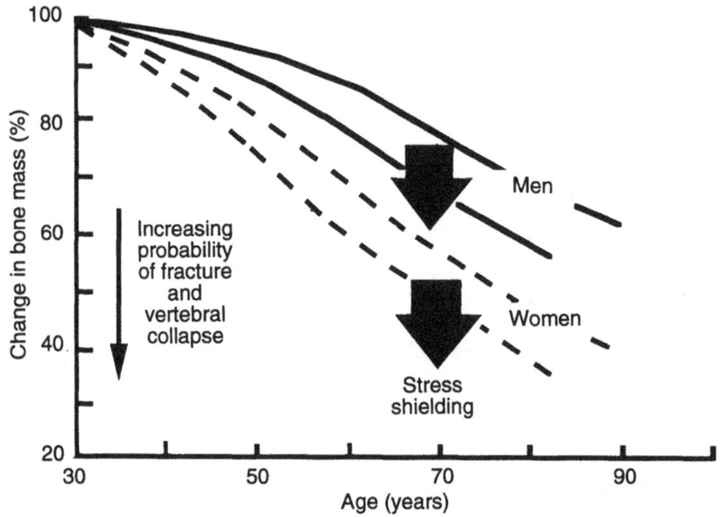

Fig. 1.1 Effect of age on bone mass. Data deterived from G.R. Mundy (1994) *Nature* **367**, 216-17.

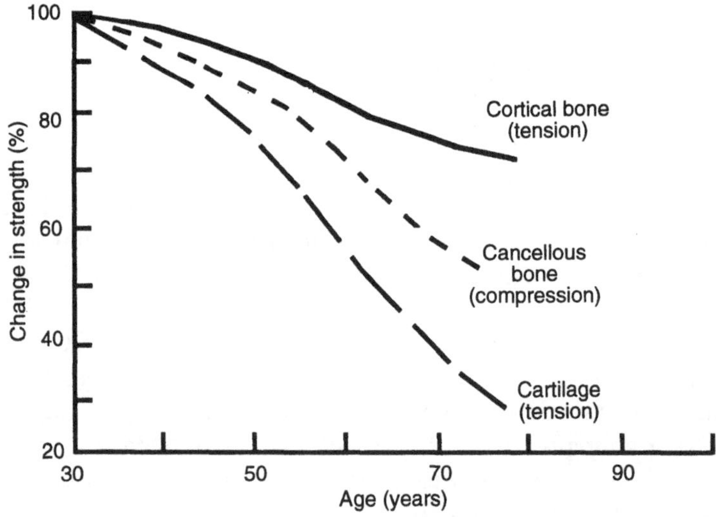

Fig. 1.2 Effect of age on strength of bone and cartilage. Data from H. Yamada (1970), *Strength of Biological Materials,* Williams and Wilkins, Baltimore, MD.

use of alloplastic (nonliving) prostheses or implants to repair, augment or replace skeletal tissues.

For many years the guiding principle used in biomaterials' and prostheses development was that the materials should be as chemically inert as possible (Simon, 1994; Cameron, 1994; Davies, 1991; Hench and Ethridge, 1982). Body fluids are highly corrosive saline solutions. The first materials used in skeletal repair were metals optimized for strength and corrosion resistance. Metallic implants for orthopaedic applications have been very successful with hundreds of thousands being implanted annually. The original applications were as removable devices, such as those for stabilization of fractures. Use as permanent joint replacements began in the 1960s with Professor Charnley's use of self-curing polymethylmethacrylate (PMMA) 'bone cement' which provided a stable mechanical anchor for a metallic prosthesis in its bony bed. This type of anchoring of implants to bone, discussed in Chapter 2, is called 'cement fixation' if PMMA cement is used (Table 1.1) (Charnley, 1972). High levels of clinical success of cemented orthopaedic implants have led to rapid growth in the use of implants, especially for hip (Chapter 2) and knee replacements (Chapter 6). The increase in the number of implants has been accompanied by an increase in the life expectancy of patients and a decrease in the average age of patients receiving an implant. This means that a growing proportion of patients will outlive the expected lifetime of their prostheses. When an implant fails, revision surgery is required. The patient, now 5-20 years older, has an increased probability of operative and postoperative complications.

POTENTIAL FOR FAILURE: A NATURAL CONSEQUENCE OF AN UNNATURAL INTERFACE

For most implants, failure originates at the interface between the biomaterial and its host tissue (Simon, 1994; Cameron, 1994; Davies, 1991; Hench and Ethridge, 1982; Hench and Wilson, 1993). The conditions for failure vary greatly depending upon the type of implant. The survivability of total hip prostheses on the one hand is generally very good, but on the other hand the percentage of middle ear aeration tubes remaining in children after only 2 to 3 years is nearly zero (Chapter 13). In some children loss of the tubes causes no long term problem, in others permanent scar tissue can impair hearing.

Table 1.1 Types of tissue attachment to skeletal prostheses

Type of Biomaterial	*Type of Attachment*	*Example*	*Typical Use*
Nearly Inert	Dense, nonporous nearly inert materials attach by bone growth into surface irregularities by cementing the device into the tissues, or by press-fitting or screwing into a defect. (Termed Morphological or Cement Fixation)	Al_2O_3 (Single Crystal and Polycrystalline)	Ball of hip replacement
		316L Stainless Steel	Joint replacement
		Co-Cr-Mo Alloy	Joint replacement
		Ti and Ti Alloy	Tooth implant
		Ultrahigh MW Polyethylene	Socket of hip replacement
Porous	For porous inert implants bone ingrowth occurs, which mechanically attaches the bone to the material. (Termed Biological Fixation)	Porous Metal Coatings	Fixation of total joint replacement
		Hydroxyapatite Coated Porous Metals	Fixation of total joints
		Hydroxyapatite	Bone replacement
Bioactive	Dense, nonporous surface-reactive ceramics, glasses, and glass-ceramics attach directly by chemical bonding with the bone. (Termed Bioactive Fixation)	Bioactive Glasses	Alveolar ridge implant, middle ear implants
		Bioactive Glass-Ceramics	Vertebral replacement
		Hydroxyapatite	Bone repair

Two factors contribute to interfacial failure. The factors are biomechanical and biochemical in nature and depend on the type of tissue being replaced. It is difficult to prevent motion at the interface of devices used to replace parts of the skeletal system because cyclic mechanical loads are transferred through those tissues. Joint replacements are subjected to interfacial cyclic shear stresses. Interfacial bonds between dissimilar materials are generally weakest in shear. It is difficult to achieve long term interfacial stability between bones and implants because of the shear stresses at the interface. Extensive wear can also occur in the articulating surfaces of prosthetic joints (Simon, 1994, Cameron, 1994). The wear of ultrahigh molecular weight polyethylene (UHMWPE) used in joint prostheses, changes local stress distributions and accelerates interfacial failure. Polyethylene wear debris migrates into tissues and causes additional damage to bone as a consequence of chronic inflammation (Revell, 1986). The combination of wear and deterioration of the bone-implant interface can be catastrophic and is the cause of the largest proportion of failed orthopaedic implants. These failure modes are discussed in Chapter 2.

Bone loss (lysis) at the interface of orthopaedic implants occurs during the process of aseptic loosening of devices (Simon, 1994; Cameron, 1994; Revell, 1986). Recent evidence also indicates that bone lysis can occur at the interface of even stable, well-fixed devices (Al-Saffar, Kadoya and Revell, 1994; Maloney *et al.,* 1990; Pierson and Harris, 1993). In either situation, inflammatory events at the bone-implant interface are linked to the loss of bone (Goldring *et al.,* 1986). The biochemical factors leading to bone lysis are exacerbated by biomechanical factors at the implant-bone interface.

Most prostheses in contact with either cortical or cancellous bone give rise to a biomechanical mismatch at their interface. Figure 1.3 compares the range of modulus of elasticity (or stiffness) of cortical bone and cancellous bone (Hench and Wilson, 1993). Values for cortical bone are between 7 and 25 GPa, depending upon age and type of bone, and between 0.05 and 0.5 GPa for cancellous bone. The composite structure of cortical bone results in a high fracture toughness (K_{1C}=2 to 12 MPa $m^{-1/2}$), and a high strain to failure (1 to 3%) which is important with respect to implant interfacial behavior.

Cancellous bone is less dense than cortical bone and consequently has a lower modulus of elasticity and a higher strain to failure (5 to 7%). The difference in stiffness between the two types of natural bone tissue ensures

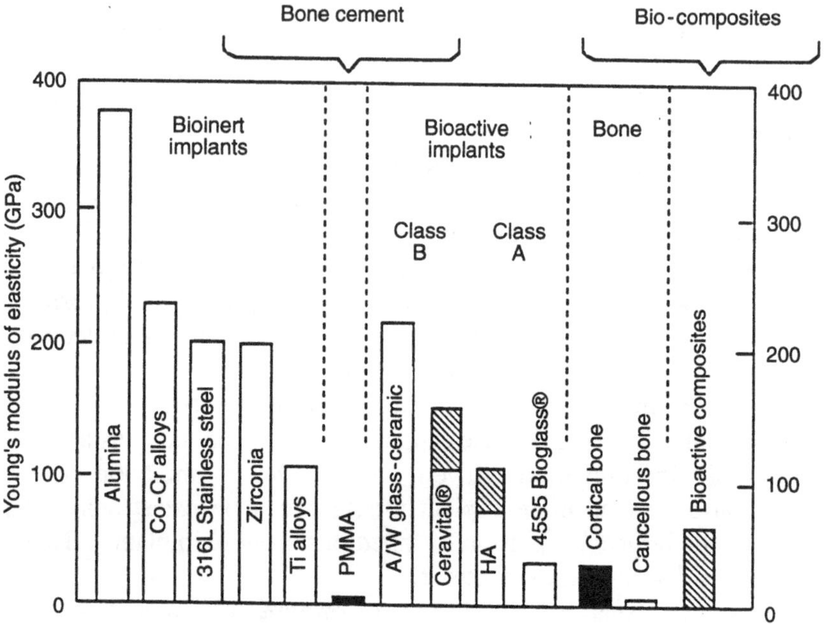

Fig. 1.3 Young's modulus of elasticity of bone compared with implant materials.

that a gradient in mechanical stress will occur across a bone when it is loaded by tendons and ligaments (Simon, 1994). However, it is impossible for most prosthetic materials in use today to produce similar biomechanically stable gradients of stiffness between an implant and its host tissues. Aging magnifies the biomechanical problem.

Figure 1.3 also shows the range of moduli of elasticity of various materials used for prostheses. There are large differences between the elastic moduli of bone and those currently used orthopedic biomaterials. For example, the mismatch in stiffness between bone and metallic implants is 10 to 20X depending upon location and quality of bone. The mismatch of moduli with cancellous bone is more than 100X.

The problem with a mismatch in elastic modulus across an interface is that the high modulus implant will carry most of the load. Thus, the bone will be 'stress shielded'. This is undesirable because living bone must be under some tensile load to remain healthy (Revell, 1986). Bone that is unloaded or is loaded in compression will undergo a biological change that leads to resorption and weakening (Simon, 1994; Cameron, 1994; Davies,

1991; Revell, 1986). This stress induced resorption accelerates the natural aging process, as illustrated in Fig. 1.1. The interface between a stress shielded bone and an implant will deteriorate as the bone structure is weakened. Loosening and/or fracture of the bone, the interface, or the implant can result. Wear debris at the interface will accelerate the weakening of stress shielded bone due to increased cellular activity involved in removing the foreign material. (Revell, 1986; Al-Saffar, Kadoya, Revell, 1994; Maloney *et al.*, 1990; Pierson and Harris, 1993; Goldring *et al.*, 1986). Use of PMMA bone cement also results in death of bone at the interface due to an exothermic reaction during polymerization. The dead bone is less elastic and weaker than living bone and is easier to fracture. Cellular remodeling of the dead bone also occurs which leads to additional weakening of the implant-bone interface (Revell, 1986).

ALTERNATIVE PROSTHETIC SYSTEMS

The potential for eventual failure of skeletal prostheses has been recognized for many years and enormous effort has been devoted to developing alternative biomedical materials and prosthetic designs that will minimize the causes of failure.

Ceramics, especially oxides, eliminate the problem of metallic corrosion of implants or structural degradation of polymers such as UHMWPE. Medical grade aluminum oxide (alumina), with high purity and small grain size has been developed for use in load bearing orthopaedic prostheses, especially as the head or ball of a total hip joint. The very high resistance to wear of alumina and low friction make it a favorable alternative to metals in this application, as described in Chapter 4. The use of alumina-alumina articulating surfaces in total hip replacement offers the best alternative for eliminating polyethylene wear debris and is widely used in Europe for younger (<50 yrs) patients (Boutin, 1987).

One approach to improved interfacial stability is the use of a porous implant or porous coating on an implant. Bone will grow into pores of >100 μm diameter and maintain the blood supply necessary for cell maintenance (Hulbert *et al.*, 1974; Holmes *et al.*, 1984). Soft tissues will grow into pores of smaller diameter. This method of fixation is often called *biological fixation* (Table 1.1). Bioactive coatings, such as hydroxyapatite (HA), applied to porous coated metallic prostheses can enhance bone growth into the pores (Ducheyne *et al.*, 1980). Chapter 3 discusses the clinical performance of these 'non-cemented' alternatives.

A third pathway to the achievement of a stable implant-tissue interface is use of bioactive fixation (Table 1.1). *A bioactive material is one that elicits a specific biological response at the interface of the material which results in the formation of a bond between the tissues and the material* (Hench and Wilson, 1993). This concept is based upon control of the surface chemistry of the material. A bioactive implant reacts chemically with body fluids in a manner that is compatible with the repair processes of the tissues. A fibrous capsule is prevented from forming by the adhesion of repairing bone (Hench *et al.*, 1971). Collagen fibrils bind within a hydroxy-carbonate-apatite (HCA) layer that grows on the bioactive implant. Bioactive implants can be used to transfer a load across an implant-tissue interface with gradients equivalent to normal physiological interfaces. However, the mechanical properties of most bioactive materials have, at present, limited their use to non-load bearing devices, discussed in Chapters 14 and 16.

THE NEED FOR COMPARATIVE CLINICAL DATA

The quest for improved clinical lifetimes of prostheses has resulted in the implantation of many different combinations of materials and designs, i.e., different prosthetic systems. Establishing the relative merits of the different systems in clinical use, the goal of this book, is important. However, it is also difficult. Many factors contribute to the difficulty: 1) Clinical studies typically report data on only one system and often on only a small group of patients. Thus, it is difficult to judge the relative merits of new systems. 2) Differences in clinical success often do not appear until 5 to 10 years have elapsed. 3) Evaluation methods differ between clinical investigators, which makes comparison of studies difficult. 4) Variation in surgical techniques often occur when new systems are used. Separating the effects of techniques on clinical success, from the prosthetic system used, is often impossible. 5) There is a learning curve associated with any new system which can affect the relative success for the early portion of the patient population. 6) Expanded use of a system to a larger number of clinics leads naturally to more variation in results.

APPROACH TAKEN

The above factors affecting clinical performance of skeletal prostheses were evaluated in a postgraduate course, 'Materials for Skeletal Prostheses', at the University of Florida, Gainesville, Florida during the Fall Semester, 1994. The course also discussed the biomechanics of the skeletal system, the alternative materials used for prostheses, designs of prostheses, mechanisms of interfacial attachment, and sources of implant failure. The 36 students were PhD candidates in materials science and engineering, health care professionals, clinicians, and government and industrial laboratory scientists. During the semester, teams of two to three students reviewed published clinical data on a particular prosthesis system and produced a basis for comparison of the clinical studies. When sufficient data were available, at comparable time periods, statistical comparisons were made leading to Kaplan-Meir survivability projections. Chapters 2 to 15 are edited versions of the student analyses. Chapter 16 is an equivalent effort conducted by the engineering and clinical team which evaluated two types of bioactive implants used for maintenance of edentulous alveolar ridges.

Our goal throughout this volume was to be as comprehensive and objective as possible in the comparison of various systems used for skeletal repair. However, there will undoubtedly be studies missed or comparisons questioned. We recognize that data on several important types of prostheses are not in this volume, for example temporomandibular joint (TMJ), maxillofacial, cranial and finger joint prostheses. We hope that any errors which appear are those of omission, which can be put right in later editions. We welcome input from any groups who are able to update the information in this volume with more recently published clinical data.

We hope that these analyses will encourage reporting of clinical studies with a standardized basis for comparison.

REFERENCES

Al-Saffar, N., Kadoya, Y. and Revell, P. (1994) *J. Maters. Sci. Maters. in Med.*, **5**, 813-818.

Boutin, P.M. (1987) *Ceramics in Clinical Applications*, (ed P. Vincenzini), Elsevier, New York, pp. 297.

Cameron, H.V. (1994) *Bone-Implant Interface,* Mosby-Year Book, Inc., St. Louis, Missouri.

Charnley, J. (1972) *J. Bone Joint Surgery*, **54B**, 61.

Davies, J.E., ed., (1991) *The Bone-Biomaterial Interface,* University of Toronto Press.

Ducheyne, P., Hench, L.L., Kagan, A., *et al.* (1980) *J. Biomed. Maters. Res.* **14**, 225.

Goldring, S.R., Jasty, M., Roelke, M., *et al.* (1986) *Arthritis Rheum.*, **29**, 836.

Hench, L.L. and Ethridge, E.C. (1982) *Biomaterials: An Interfacial Approach*, Academic Press, New York.

Hench, L.L., Splinter, R.J., Allen, W.C., *et al.* (1971) *J. Biomed. Maters. Res.,* **2** (1), 117-41.

Hench, L.L. and Wilson, J. (1993) *Introduction to Bioceramics*, World Scientific Publishing Co., London-Singapore.

Holmes, R.E., Mooney, R.W. and Bucholz, R.W. (1984) *Clin. Orthop. Relat. Res.*, **188**, 282-92.

Hulbert, S.F., Matthews, J.R. and Klawitter, J.J. (1974) *Biomed. Maters. Symp.* **5**, 85-97.

Maloney, W.J., Jasty, M.J., Rosenberg, A., *et al.* (1990) *J. Bone Joint Surgery*, **72B**, 966.

Pierson, J.L. and Harris, W.H. (1993) *J. Bone Joint Surgery*, **75A**, 268.

Revell, P. (1986) *Pathology of Bone*, Springer-Verlag, Berlin.

Simon, S.R., ed. (1994) *Orthopaedic Basic Science,* American Academy of Orthopaedic Surgeons, Rosemond, IL.

2

Low-Friction Total Hip Arthroplasties

Jason A. Griggs
Thomas J. Hill
Shawn H. Gallagher
Teresa Schimmel

INTRODUCTION

Total hip arthroplasty (THA) is one of the most frequently performed operations in orthopaedics. Three decades of clinical studies have resulted in a large body of data regarding the survivorship, performance, and revision of THAs. The benefits of compiling these data into a single reference lie in aiding clinicians in decision making and patient education. It is important to know whether the life expectancy of current prosthetic devices may result in device failure at the time when revision surgery may produce a high risk to the patient.

There are several difficulties in combining clinical studies. The variations in technique between surgeons have such profound effects that comparison of studies is uncertain. Often the literature does not contain the specific information needed to compare results. The data concerning gender, weight, activity, age, and preoperative condition of patients is only occasionally provided and is rarely quantitative. At least 17 different methods of assessing the postoperative performance of THAs have been used.

These uncertainties make it difficult to combine data from different studies with confidence. However, there is a need to assess the general trends of survivability of THA. It is the purpose of this chapter to review the current THA literature with consideration given to the age and gender of patients.

The Charnley (1972) low-friction THA was the first systematic effort to engineer a THA system with reduced torque and wear. This was achieved through: 1) choosing a surface with lubrication characteristics similar to those of a natural joint, 2) optimizing geometry and mechanical properties to minimize and evenly distribute loads, and 3) providing a method of fixation for rigid force transmission to the surrounding bone.

Charnley (1979) felt a well lubricated joint to be a priority because slow debris production and small debris size prevent the chronic inflammation that causes bone resorption. Nature seems to have adopted a mixture of the principles of hydrodynamic lubrication and boundary lubrication. Hydrodynamic lubrication utilizes the convergence of wedge-shaped fluid layers to generate a separating pressure between two surfaces under the action of their rotation. Boundary lubrication involves the chemical bonding of a thin layer of fluid to produce an easily sheared cladding for each surface. Charnley's first choices for prosthetic surfaces were materials that felt slippery in air, such as polytetrafluoroethylene (PTFE). Pure PTFE and glass-reinforced PTFE proved to have poor wear properties that resulted in disastrous complications. In searching for a substitute material, Charnley (1979) tested combinations of stainless steel against polyethylene, cobalt-chrome against polyester, alumina against alumina, cobalt-chrome against cobalt-chrome, and alumina against polyethylene. The stainless steel against ultrahigh molecular weight polyethylene (UHMWPE) combination has a coefficient of friction roughly five times that of the PTFE system, but UHMWPE surfaces are easily wetted by synovial fluid. This allows hydrodynamic and boundary lubrication to reduce wear rates by 500 to 1000 times *in vivo*. For this reason, UHMWPE became the material of choice for the acetabular component (Fig 2.1).

The primary geometric variable in the acetabular component is the size of the socket which holds the femoral head. Low-friction arthroplasty is characterized by a combination of a small femoral head with a socket of maximum external diameter to produce maximum wall thickness. Charnley (1979) successively reduced the femoral head diameter to arrive at an optimum 22 mm. He used stress polariscopy to determine that a thick wall delivers more evenly distributed loads to the acetabulum with no stress concentrations to cause bone resorption. In addition, a smaller femoral head has less surface area and requires less force against the fibrous capsule to

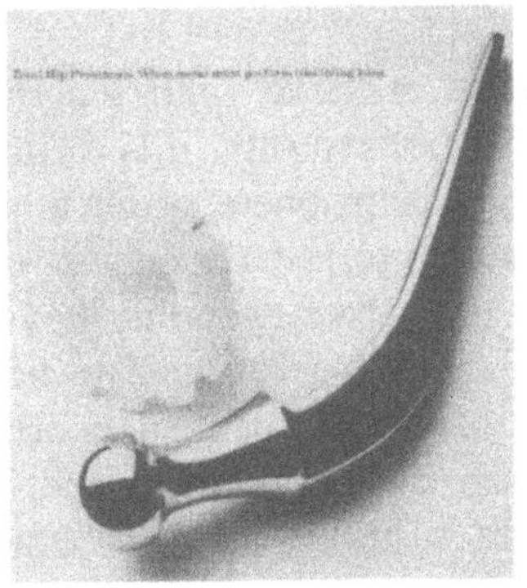

Fig 2.1 A Charnley total hip with a stainless steel stem and UHMWPE acetabulum (Zimmer Corp.)

produce momentary dislocation during hyperextension. This was originally viewed as a negative aspect as this design seemed prone to frequent dislocation; however, it was later realized that this meant smaller forces are transferred to the cement layer, producing less loosening. Eventually a long posterior wall was added to the cup geometry to prevent most postoperative dislocations.

The geometry and mechanics of the femoral component have improved through changes in many variables. Charnley elected not to switch to more fatigue resistant alloys, such as cobalt-chrome and titanium, as a solution to the initial problem of femoral neck fracture. He saw this as a partial solution because the motion would continue and eventually damage the cement layer (Charnley, 1979). Instead, he chose to decrease the amplitude of the deflection by maximizing the cross-sectional area of the stem and avoiding sharp, stress-raising features on the surface. The increased cross-section can fit into the medullary canal by removal of the trochanter. Also, a reduction in offset reduced the bending moment and stresses on the cement layer between the prosthesis concavity and the medial femoral neck.

Charnley was the first to use self-curing cement in the medullary canal instead of just on the upper portions of the shaft. His addition of the flanged 'cobra' stem maximizes the shaft surface area in contact with cement, providing a more even distribution of forces along the cement layer. It also contacts the inside of the medullary canal, centering the shaft and aiding the surgeon in avoiding motion of the shaft during cement curing. Plugs are now inserted in the distal end of the femoral canal prior to cement injection, and cement guns are used to increase the injection pressure.

Attempts to improve attachment of the acetabular component brought about many variations in design. Originally, fixation through fibrous ingrowth was attempted, but the use of cement was adopted once the success of cement in the medullary canal became obvious. The original acetabular shape was a hemispherical cup. In order to preserve limb length, any deepening of the acetabular socket must occur transversely. It is difficult to achieve a high injection pressure by inserting a hemisphere at this angle. In the pressure injection cup, a lobe was added both to prevent cement escape and to contact the edge of the acetabulum, aiding the surgeon in preventing motion during cement curing. Press-fit implants were designed to be used without cement because chronic inflammation due to cement debris was thought to be a factor in acetabular loosening. These non-cemented THAs loosened faster than the Charnley prostheses because the presence of polyethylene debris is actually the controlling factor behind chronic inflammation.

The current advances in THA technology lie in developing more modular components (Evans *et al.*, 1993). The mixability of modular components allows polyethylene liners to be placed in metal acetabular cups. The metal cup can remain fixed while only the worn liner is replaced. For the femoral component, modularity means an ability to adjust limb length and femoral offset by choosing between heads. Also, the head can be removed during revision surgery to improve access to the acetabular components. There are drawbacks to these designs that have prevented them from becoming popular. The screw holes, that allow fastening of the polyethylene to the metal in the acetabular component, channel wear debris behind the acetabular component where it does the most damage. Modular femoral heads have been known to separate from the stem in the case of a joint dislocation (Pellicci and Haas, 1990). The future of THA implant design lies in alleviating these obstacles to allow wide use of modular designs.

ASSESSMENT METHODS

Several different methods of assessing the success or failure of THAs exist in the literature. In order to compare the results of separate studies, it is important to understand the differences in their methods of evaluation. The Charnley (d'Aubigne and Postel, 1954), Harris (1969), and Mayo (Kavanagh and Fitzgerald, 1985) hip scores are the most prominent scoring systems.

The Charnley or Wrightington hip score, is noted for its simplicity. A short questionnaire asks patients to assign a number between 1 and 6 to pain, movement, and walking (Table 2.1). The mean of these three values is the

Table 2.1 Charnley method for postoperative assessment of total hip arthroplasties (6 points*)

	Score
Pain (6 points)	
No pain	6
Slight or intermittent, pain on starting to walk but getting less with normal activity	5
Only after some activity, disappears quickly with rest	4
Tolerable, permitting limited activity	3
Severe on attempting to walk, prevents all activity	2
Severe and spontaneous	1
Movement (6 points)	
260°	6
210°	5
160°	4
100°	3
60°	2
0-30°	1
Walking (6 points)	
Normal	6
No stick but a limp	5
Long distances with one stick, limited without a stick	4
Limited with one stick (less than one hour), difficult without a stick, able to stand long periods	3
Time and distance very limited with or without sticks	2
Few yards or bedridden, two sticks or crutches	1

*Total score is the mean of the three main categories

over-all score. Published results are often described as excellent (5-6), good (4-5), fair (2-4), or poor (0-2), excellent and good are considered satisfactory.

The Harris hip score was developed in order to integrate assessments of function, motion, and quality of life (see Table 2.2).

The Mayo hip score includes radiographic assessment. This scoring system is split into an 80 point clinical assessment (Table 2.3) and a 20 point radiographic assessment (Table 2.4).

The Harris hip score is based on a possible 100 points broken down as follows 44 points for pain, 47 points for function, 5 points for range of motion, and 4 points for absence of deformity. The scoring for pain and function involves the patient filling out a questionnaire. The patients are asked to assess their ability to perform daily activities such as climbing

Table 2.2 Harris method for postoperative assessment of THA (100 points)

	Score
Pain (44 points)	
None or ignores it	44
Slight, occasional, no compromise activities	40
Mild pain, no effect on average activities, rarely may have moderate pain with unusual activity, may take aspirin	30
Moderate pain, tolerable but makes concessions to pain, some limitation of ordinary activity or work, occasional pain medication stronger than aspirin	20
Marked pain, serious limitation of activities	
Totally disabled, crippled, pain in bed, bedridden	10
Function (47 points)	0
Gait (33 points)	
Limp (11 points)	
None	11
Slight	8
Moderate	5
Severe	0
Support (11 points)	
None	11
Cane for long walks	7
Cane most of the time	5
One crutch	3
Two canes	2
Two crutches	0
Not able to walk	0
Distance (11 points)	
Unlimited	11
Six blocks	8
Two or three blocks	5
Indoors only	2
Bed and chair	0
Activities (14 points)	
Stairs (4 points)	
Normally without using railing	4
Normally using railing	2
In any manner	1
Unable to do stairs	0
Shoes and socks (4 points)	
With ease	4
With difficulty	2
Unable	0
Sitting (5 points)	
Comfortably in ordinary chair for one hour	5
On a high chair for one-half hour	3
Unable to sit comfortably in any chair	0
Enter public transportation (1 point)	
Absence of deformity (4 points)	
Range of motion (5 points)	

Table 2.3 Mayo method for postoperative clinical assessment of total hip arthroplasties (80 points)

	Score
Pain (40 points)	
None	40
Slight or occasional	35
Moderate	20
Severe	0
Function (20 points)	
Distanced walked (15 points)	
Ten blocks or more	15
Six or more blocks	12
One to three blocks	7
Indoors	2
Unable to walk	0
Supports aids (5 points)	
None	5
Cane occasionally	4
Cane or crutch full-time	3
Walker	1
Unable to walk	0
Mobility and muscle power (20 points)	
Car (5 points)	
With ease	5
With difficulty	3
Unable	0
Foot care (5 points)	
With ease	5
With difficulty	3
Unable	0
Limp (5 points)	
None	5
Slight	3
Severe	0
Stairs (5 points)	
Normal	5
With rail	4
One step at a time	2
Unable	0

Table 2.4 Mayo method for postoperative radiographic assessment of total hip arthroplasties (20 points)

Acetabulum (10 points)	Score
Incomplete bone-cement lucent line	10
Complete line since surgery ≤1 mm	8
Progressive line since surgery ≤1 mm	7
Complete or progressive line >1 mm in any one zone	4
Component migration	0
Femur (10 points)	
Incomplete bone-cement lucent line	10
Complete line since surgery ≤1 mm	8
Progressive line since surgery ≤1 mm	7
Complete or progressive line >1 mm in any one zone	4
Subsidence:	
≤2 mm	4
>2 mm	0
Prosthesis-cement lucent line:	
≤1 mm	4
1-2 mm	2
>2 mm	0

stairs, using public transportation, sitting, and putting on shoes and socks. In scoring the range of motion, the extreme ranges of arc are weighted less than the central range of arc. Lack of deformity is judged on the presence of a permanent flexion or rotation or the presence of a difference in limb length. The clinical section is composed of 40 points for pain, 20 points for function, and 20 points for mobility and muscle power. Patients grade themselves on activities such as entering cars, foot care, and using stairs. The radiographic section is broken down into equal points for the acetabulum and the femur. Radiographs are graded on the presence and progression of lucent lines. The Mayo score is generally lower than the Harris score for a given THA (Kavanagh and Fitzgerald, 1985) because the former incorporates radiographic data and the latter is purely clinical. Radiographic assessments frequently find loosening in THAs that never develop any clinical problems. No study to date has found a significant correlation between radiographic evidence and clinical performance (McCoy *et al.*, 1988; Wykman *et al.*, 1988; Wykman, Selvik and Goldie, 1988; Gudmundsson, Harving and

Pilgaard, 1989; Stringa *et al.*, 1989; Wroblewski, Taylor and Siney, 1992; Boeree and Bannister, 1993)

COMBINED SCORE

In an effort to relate these scoring methods for this chapter, the basic areas of measurement were compared on a value basis. The Harris hip score was used as the standard because the greater level of detail involved permits a better match for each category than the other methods. Table 2.5 lists formulas for converting scores from the Charnley six-point scale and the Mayo 80-point clinical scale to the Harris 100-point scale. The Charnley categories of pain and walking are similar to the Harris categories of pain and gait, respectively. The Charnley category of movement applies to Harris categories of activity, absence of deformation, and range of motion. These category scores are multiplied by weighting factors, resulting in a 100-point total. The radiographic assessment is not considered for conversion from Mayo scores because the Harris score contains no comparable categories. These data can be safely disregarded because they have shown no correlation with clinical performance. The Mayo categories of 'pain', 'limp', support aids, 'distance walked', and 'stairs' are comparable to the Harris categories of the same names. The Mayo category of 'foot care' applies to both the Harris categories of 'shoes' and 'range of motion'. The Mayo category of car use applies to both the Harris categories of 'sitting' and 'entering public transportation'. Absence of deformity is determined by the lack of a limp.

In addition to using different scoring methods, clinical studies make use of a variety of definitions to assign success or failure to a particular hip. Common criteria for success include a good or excellent score, a lack of pain, or greater than preoperative functionality. Here, for the compilation of success rates from differing studies, *the definition adopted for success is the lack of a revision operation during the follow-up period in question.*

FAILURE MODES

The femur can be divided into seven zones for radiographic failure analysis as shown in Fig 2.2. Proximal radiographs will show lucent lines where bone cement has failed and when loosening is present. Looseness is defined as

Table 2.5 Method for converting Charnley and Mayo hip scores to the equivalent Harris hip scores

Harris	Charnley	Mayo
Pain	7.33 (pain)	Pain
Function		
Gait	5.50 (Walking)	
Limp		2 (Limp)
Support		Support aids
Distance		Distance walked
Activities	2.33 (Movement)	
Stairs		Stairs
Shoes and socks		Foot care
Sitting		Car use
Enter public transportation		Car use
Absence of deformity	0.33 (Movement)	Limp
Range of motion	0.83 (Movement)	Foot care

fractured cement and an interface gap such as a radiolucent zone at the stem-cement or the cement-bone interface (Gruen, McNeice and Amstutz, 1979).

The results of loosening are wear and migration. Wear is the chief obstacle to prolonging the lifetime of THAs because the chronic inflammation associated with wear debris results in bone resorption. The change in loading geometry produced by migration is expected to have profound mechanical effects; however, a large degree of migration is generally necessary to require revision surgery. Radiographic examination can still be considered a valuable tool because it allows the clinician to monitor progressing migration through noninvasive means.

Figure 2.3 depicts the classification of several common modes of failure. Mode Ia is piston action of the metal stem within the cement sheath. This mode is characterized by a radiolucent line appearing in zone 1, usually accompanied by the punching-out of cement at the distal end of the stem. Mode Ib is the piston action of an embedded stem within the bone. When this occurs, a radiolucent line is generally present around most of the cement-bone interface. Lack of calcar support causes medial midstem pivoting (Mode II) which causes a radiolucent line in zone 1 along with cement cracking in zones 2 and 6. Lack of zone 3 support causes calcar

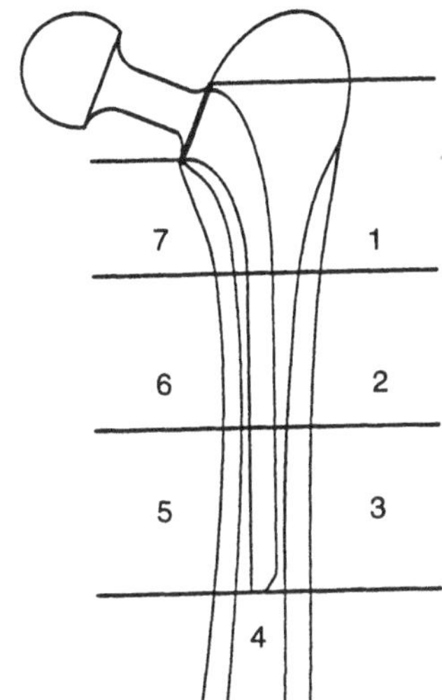

Fig. 2.2 Regions of failure.

Fig. 2.3 Classification of failure.

Mode		Description	
I	Ia	Pistoning: stem with cement	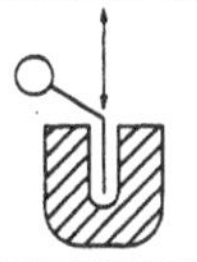
	Ib	Pistoning: stem with bone	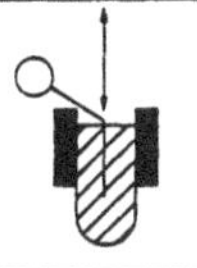
II		Medial midstream pivot	
III		Calcar pivot	
IV		Bending cantilever (fatigue)	

pivoting (Mode III) which occurs as a medial-lateral motion in the distal end of the embedded stem. Mode IV is a cantilever action which results in fatigue of the stem neck. It is characterized by a radiolucent line in zone 6 and usually zone 2 while zone 4 remains firmly cemented.

RESULTS

Recently, many long term results for Charnley low-friction arthroplasty have been reported which show the survivability of this prosthesis for up to a 20-year period. (Table 2.6) These results reconfirm the success that Charnley and Cupic (1973) first reported for a 9 to 10-year follow-up. Since that original evaluation of the prosthesis, more recent reports have looked not only at the success rate of the prosthesis but also at other factors which affect its performance such as gender, age and weight of the patient; bilateral versus unilateral disease; and cemented versus non-cemented fixation.

Table 2.6 Results from long-term studies reporting the survivorship of low friction total hip arthroplasties

Study	*Followed Hips*	*Total Revisions*	*Period (yr)*
Eftekar (1971)	138	6	7-8
Charnley (1972	210	10	4-5
Charnley and Cupic (1973)	106	7	9-10
Charnley (1975)	6500	17	3.5
Maier *et al.* (1977)	95	11	4-7
Beckenbaug and Ilstrup (1978)	301	23	4-7
Charnley (1979)	77	9	12-15
Cupic (1979)	409	3	11.5
Olsson *et al.* (1981)	151	14	5-10
Salvati *et al.* (1981)	67	3	9.5-11.5
Stauffer (1982)	231	17	10
Collis (1988)	525	37	10
Collis (1988)	156	27	12.1
Sudmann *et al.* (1983)	113	1	5-8
Brady and McCutchen (1986)	170	15	10
Dall *et al.* (1986)	98	14	12
Eftekar and Tzitzikalakis (1986)	696	31	5-15
Older (1986)	153	9	10-12
Terayama (1986)	107	6	5
Terayama (1986)	11	3	10
Wroblewski (1986)	116	12	15-21
Welch *et al.* (1988)	65	10	15-17
McCoy *et al.* (1988)	40	5	15-3
Russotti *et al.* (1988)	215	0	5-7
Wejker and Stenport (1988)	248	20	5
Wejker and Stenport (1988)	163	13	10
Kavanagh *et al.* (1989)	166	37	15
Salvati *et al.* (1989)	40	12	14.4-16.2
Mulroy and Harris (1990)	105	7	10-12.7
Hozack *et al.* (1990)	1041	83	10
Solomon *et al.* (1992)	156	14	10
Stringa *et al.* (1992)	76	8	12
Wroblewski *et al.* (1992)	57	0	19-25
Wroblewski and Siney (1992)	1342	141	10.3
Boeree and Bannister (1993)	46	1	7
Boeree and Bannister (1993)	46	5	10
Boeree and Bannister (1993)	46	6	12
Schulte *et al.* (1993)	322	32	20
Izquierdo and Northmore-Ball (1994)	148	7	10
Kavanagh *et al.* (1994)	112	38	20
Neumann *et al.* (1994)	103	9	15-20
Sullivan *et al.* (1994)	84	13	16-22

In 1977, Maier and associates reported 83% of 95 hip arthroplasties in good or excellent clinical condition over 4 to 7 years. The majority of the non-satisfactory prostheses (11.2% of total) were revised due to aseptic loosening. Another 3.5% of the hips showed both clinical and radiographic loosening but were still in use while 4.9% demonstrated radiographic evidence but no clinical signs.

In Denmark, Olsson *et al.* (1981) reported a 79% clinical success (free from pain) rate in 151 hips where 14 arthroplasties were revised. These patients were followed for 5 to 10 years. Olsson noted that loosening had occurred in over 50% of the patients over 80 kg and younger than 60 years of age.

At the Hospital for Special Surgery, Salvati *et al.* (1981) studied 100 Charnley hip prostheses. Of the 67 hips which could be evaluated after 10 years, 59 demonstrated good (22) or excellent (37) results according to the Hospital for Special Surgery scoring system. Of the 54 hips which were radiographically evaluated, 23 showed lucencies that appeared to have no significant bearing on clinical performance.

Collis (1988) reported 8 to 10-year clinical results with the radiographic findings for 350 consecutive THAs. Thirteen femoral components and one acetabular component required revision. Of the 129 hips that could be assessed after 10 years, heterotopic bone formation, calcar resorption, and distal cortical hypertrophy all appeared to be radiographic findings with little effect on clinical performance. In this study, the 29 patients who were under 50 years of age had a revision rate not significantly different from the group as a whole.

In 1983, Sudmann *et al.*, performed a 5 to 8-year study comparing the Charnley THA to the Christiansen THA. For 113 hips in 87 patients, only one Charnley prosthesis failed during the study, while the Christiansen prostheses failed 19 times in 90 hips of 81 patients. Thus the Charnley THA demonstrated a 99% success rate compared with a success rate of 79% for the Christiansen THA.

Gudmundsson *et al,* (1985) examined 125 patients both clinically and radiographically for mechanical loosening after hip replacements. After ten years, they reported that 29% of the patients displayed radiographic signs of definite or probable loosening, but Gudmundsson stated 'there was a poor correlation between the clinical and radiographic results, as 86% of the hips were free of significant pain.' He also noted that, in patients under the age of 60, the loosening rate for males was four times greater than for females.

In 1986, Older, reported results of a 10- to 12-year study stating that 'low-frictional torque arthroplasty remains a sound concept of design' due to a satisfactory rating in 88% of the THAs with only 6% requiring revision. Two percent were revised due to aseptic loosening.

Also in 1986, Wroblewski reported one of the longest follow-up periods for the Charnley THA. He looked at 116 arthroplasties in 93 patients after a mean period of 16.6 years. The age of patients at operative time varied from 20 to 71 years of age. Many of these patients received the prosthesis because of osteoarthritis (52 patients) or rheumatoid arthritis (19 patients). In spite of these conditions 85.3% of the patients were 'completely pain free' and another 11.2% had 'only occasional discomfort' while 78% had full or near full range of motion. He also reported a significant correlation between time and wear but no correlation between wear and patients' weights.

A 15 to 17-year study by Welch *et al.* (1988) showed a revision rate of 16% for 65 patients. This is one of the higher revision rates reported. The major cause for revision was fracture of the femoral head, but as Welch reports, 'with modern alloys, this should not be a long term problem.'

Russotti (1988) performed a retrospective clinical and radiographic review of 251 cemented THAs that had been followed-up for a minimum of 5 years. At the follow-up examinations, 98% of the patients had excellent results. Using the Harris scoring system, the patients had a 47 preoperative score and a 97 postoperative score.

In 1989, Salvati *et al.*, tested 40 surviving hip prostheses. After a period of 14.4 years to 16.2 years (15.3 years mean), 87.5% of the arthroplasties had a good or excellent rating. The failure rate of the acetabular components (4%) was half that of the femoral components (9%). Salvati reported that of the 40 THAs, 27 displayed wear in the range 0.2-8.5 mm (1.85 mm mean) and that most of these were in males, suggesting that males are a high risk group.

Hozack *et al.* (1990) at Thomas Jefferson University performed a survivorship analysis on 1,041 Charnley THAs over a 10-year period. They reported a 92% survival rate. Again acetabular survival (99%) was higher than femoral survival (96%). Hozack *et al.* concluded that for high risk groups such as active patients, male patients, patients under 50 years old, or patients weighing 170 or more pounds, a non-cemented prosthesis would be better.

Wykman *et al.* (1991) compared cemented arthroplasties and press fit (HP Garches) arthroplasties over 5 years. Seventy-five THAs of each type were studied. From these data they drew three conclusions: 1) the patients with cemented prostheses recovered faster than those with press-fit prostheses, 2) the cemented prostheses had more 'excellent' or 'good' ratings (79%) than the press-fit ones (70%), and 3) the press-fit prostheses had a higher revision rate for mechanical loosening (18.7%) than the cemented ones (6.7%).

In the last two years, longer term reports have become available. One of these was by Boeree and Bannister (1993), who used the Harris scoring system to study the survivorship of THAs in patients younger than 50 years of age over a 10- to 18-year period. The survivors' mean Harris score for the 46 hips studied was 93. This study also found no relationship between radiographic data and clinical performance.

Another recent study was performed by Schulte *et al.* (1993). They reviewed 330 arthroplasties in 262 patients with a minimum 20-year follow-up. Of the 98 hips (83 patients) still viable, 83 (85%) had no pain, 14 (14%) had mild pain, and only 1 (1 %) had moderate pain due to the prostheses. In 322 followed hips, the rate of revision due to aseptic loosening of the acetabular component was only 6% while the femoral component revision rate was 2%.

One of the most complete analyses of Charnley low-friction arthroplasties was reported by Kavanagh *et al.* (1994). Data for this same group of THAs were reported on previous occasions (Beckenbaug and Ilstrup, 1978; Kavanagh *et al.*, 1989). This investigation used Mayo scoring and the Kaplan-Meier prediction. Out of the 112 hips studied,102 (91 %) reported no or slight pain. The mean Mayo score was 79. Loosening of the acetabular component appeared to be probable in 12 hips and possible in another 33 hips while femoral loosening appeared probable in 25 hips and possible in another 14 hips. The patients' ages at implantation and the probability of failure within 20 years for the different age groups were predicted to be: age less than 59 (27% failures), 59-65 (13%), 66-70 (7.5%), and 70 or more (12%). The investigation reported that males are at a higher risk than females.

Neumann *et al.* (1994) have recently performed a 15- to 20-year study of 103 hips from 241 original arthroplasties. This showed a survival rate of 96.3%. Neumann used the Charnley scoring system to analyze the arthroplasties. Ninety-five percent were found to be good or excellent (4+) with respect to pain, 73% were found to be good or excellent (4+) with respect to function, and 93% were found to be good or excellent (4+) with respect to movement. The Kaplan-Meier prediction was for a 10.7% revision rate at 20 years.

A recent study conducted through the Norwegian Arthroplasty Register includes one of the largest patient populations studied to date with 15,335 prostheses (Havelin *et al.*, 1994). Havelin *et al.* reported unusually detailed population demographics and analyzed variations in results between differing patient groups. They found a revision rate for non-cemented THAs twice that for cemented THAs. This difference was more pronounced in patients under 60 years old, especially men. The femoral components had a revision rate about three times that of the acetabular components.

SUMMARY

Table 2.6 summarizes the clinical results of 42 published clinical studies. Kaplan-Meier prediction (1958) was used to estimate cumulative percentage of revisions over time for Charnley THAs reported in Table 2.6. The Havelin *et al.* study is excluded in order to avoid skewing the data. The predicted survival rates are plotted in Fig 2.4.

The rates at five-year intervals are 99.41 ± 0.02% at five years, 95.48 ± 0.04% at ten years, 83.12 ± 0.18% at 15 years, and 66.53 ± 0.35% at 20 years. A lifetime of 19.48 ± 0.08 years is expected for the average arthroplasty.

These estimates include data from studies in which no demographics were reported for the patient populations, so predictions should be interpreted in terms of the average patient. THAs can be expected to have shorter lifetimes in heavier individuals, males, and patients with active lifestyles.

Many of these reports support the idea that males are high failure risks (Wykman *et al.*, 1988; Wykman, Selvik and Goldie,1988; Gudmundsson, Harving and Pilgaard, 1989; Stringa *et al.*, 1989; Wroblewski, Taylor and Siney, 1992; Boeree and Bannister, 1993; Havelin *et al.*, 1994). Most studies agree that THA life expectancies are dependent on the patient's age at implantation, although some studies refute this idea (Sudmann *et al.*, 1983; Boeree and Bannister, 1993; Collis, 1988). Two ideas which appear consistently are that radiographic data and actual clinical performance are not correlated (Gudmundsson, Harving and Pilgaard, 1989; Stringa *et al.*, 1989; Wroblewski, Taylor and Siney,1992; Boeree and Bannister, 1993) and that

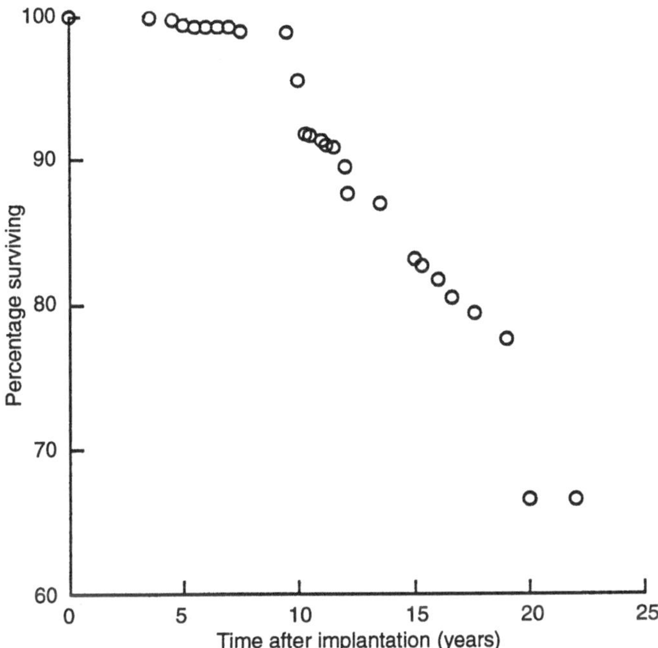

Fig 2.4 Results of a Kaplan-Meier analysis for lifetimes of Charnley THA.

femoral components are the source of more loosening and failure than acetabular components (Charnley, 1979; Wejkner and Wiege, 1987; Wejkner *et al.*, 1988; Salvati *et al.*, 1989; Stringa *et al.*, 1989; Hozack *et al.*, 1990; Havelin *et al.*, 1994; Kavanagh *et al.*, 1994). Charnley THAs are reported to have greater lifetimes than cementless systems (Sudmann *et al.*, 1983; Olsson *et al.*, 1986; Wykman *et al.*, 1988; Mattsson *et al.*, 1990; Havelin *et al.*, 1994; Onsten *et al.*, 1994).

General conclusions from these reports are that a patient's gender, weight, age, and activity level are all factors determining arthroplasties' life expectancies. It does appear clear that the Charnley THA has a low risk of failure for patients 65 and older and is the standard with which all new arthroplasties should be compared.

REFERENCES

Beckenbaug, R.D, and Ilstrup, D.M. (1978) A review of three hundred and thirty-three cases with long follow-up. *J Bone Joint Surg* **60A**(3), 306-313.

Boeree, N.R. and Bannister, G.C. (1993) Cemented total hip arthroplasty in patients younger than 50 years of age. Ten to 18 year results. *Clin Orthop* **287**, 153-159.

Brady, L.P. and McCutchen, J.W. (1986) A ten-year follow-up study of 170 Charnley total hip arthroplasties. *Clin Orthop* **211**, 51-54.

Charnley, J. (1972) The long-term results of low-friction arthroplasty of the hip performed as a primary intervention. *J Bone Joint Surg* **54B**(1), 61-76.

Charnley, J. (1975) Fracture of femoral prostheses in total hip replacement. A clinical study. *Clin Orthop* **111**, 105-120.

Charnley, J. (1979) *Low Friction Arthroplasty of the Hip.* New York: Springer-Verlag.

Charnley, J. and Cupic, Z. (1973) Results of low friction arthroplasty of the hip. *Clin Orthop* **95**, 9-25.

Collis, D.K. (1988) Long term results of an individual surgeon. *Orthop Clin North Am* **19**, 541-550.

Cupic, Z. (1979) Long-term follow-up of Charnley arthroplasty of the hip. *Clin Orthop* **141**, 28-43.

Dall, D.M., Grobbelar, C.J., Learmonth, I.D., Dall, G. (1986) Charnley low-friction arthroplasty of the hip. Long-term results in South Africa. *Clin Orthop* **211**, 85-90.

d'Aubigne, M.R., Postel, M. (1954) Functional results of hip arthroplasty with acrylic prostheses. *J Bone Joint Surg* **36A**, 451-475.

Eftekar, N. (1971) Charnley low friction torque arthroplasty. A study of long-term results. *Clin Orthop* **81**, 93-104.

Eftekar, N.S. and Tzitzikalakis, G.I. (1986) Failures and reoperations following low friction arthroplasty of the hip. *Clin Orthop* **211**, 65-78.

Evans, B.G., Salvati, E.A., Huo, M.H., Huk, O.L. (1993) The rationale for cemented total hip arthroplasty. *Orthop Clin North Am* **24**(4), 599-610.

Gruen, T.A., McNeice, G.M. and Amstutz H.C. (1979) Modes of failure of cemented stem-type femoral components. *Clin Orthop* **141**, 17-27.

Gudmundsson, G.H., Harving, S. and Pilgaard, S. (1989) The Charnley total hip arthroplasty in juvenile rheumatoid arthritis patients. *Orthopaedics* **12** (3), 385-388.

Gudmundsson, G.H., Hedeboe, J. and Kjaer, J. (1985) Mechanical loosening after hip replacement. Incidence after 10 years in 125 patients. *Acta Orthop Scand* **56** (4), 314-317.

Harris, W.H. (1969) Traumatic arthritis of the hip after dislocation and acetabular fractures: treatment by mold arthroplasty. *J Bone Joint Surg* **51A** (4), 737-755.

Havelin, L.l., Espohaug, B., Vollset, S.E., Engesaster, L.B. (1994) Early failures among 14,009 cemented and 1,326 uncemented prostheses for primary coxarthrosis. *Acta Orthop Scand* **65** (1), 1-6.

Hozack, W.J., Rothman, R.H., Booth, R.E. Jr, Balderston, R.A., Cohn, J.C., Pickens, G.T. (1990) Survivorship analysis of 1,041 Charnley total hip arthroplasties. *J Arthroplasty* **5** (1), 41-47.

Izquierdo, R.J. and Northmore-Ball, M.D. (1994). Long-term results of revision hip arthroplasty. Survival analysis with special reference to the femoral component. *J Bone Joint Surg* **76B** (1), 34-39.

Kaplan, E.L. and Meier, P. (1958) Nonparametric estimation from incomplete observations. *Am Stat Ass J* **53**, 457-481.

Kavanagh, B.F., Dewitz, M.A., llstrup, D.M., Stauffer, R.N., Coventry, M.B. (1989) Charnley total hip arthroplasty with cement. Fifteen-year results. *J Bone Joint Surg* **71A** (10), 1496-1503.

Kavanagh, B.F. and Fitzgerald, R.H. Jr (1985) Clinical and roentgenographic assessment of total hip arthroplasty. *Clin Orthop* **193**, 133-140.

Kavanagh, B.F., Wallrichs, S., Dewitz, M., Berry, D., Currier, B., llstrup, D., Coventry, M.B. (1994) Charnley low-friction arthroplasty of the hip. Twenty-year results with cement. *J Arthroplasty* **9** (3), 229-234.

Maier, S., Griss, P., Rahmfeld, T., Dinkelacker, T. (1977) 4-7 year results of total hip arthroplasty with special reference to late complications. Z *Orthop* **115** (3), 274-283.

Mattsson, E., Brostrom, L.A. and Linnarsson, D. (1990) Walking efficiency after cemented and noncemented total hip arthroplasty. *Clin Orthop* **254**, 170-179.

McCoy, T.H., Salvati, E.A., Ranawat, C.S., Wilson, P.D. Jr (1988) A fifteen-year follow-up study of one hundred Charnley low-friction arthroplasties. *Orthop Clin North Am* **19** (3), 467-476.

Mulroy, R.D. and Harris, W.H. (1990) The effect of improved cementing techniques on component loosening in total hip replacement: An eleven year readiological review. *J Bone Joint Surg* **72B**, 757-760.

Neumann, L., Freund, K.G. and Sorenson, K.H. (1994) Long-term results of Charnley total hip replacement. Review of 92 patients at 15 to 20 years. *J Bone Joint Surg* **76B** (2), 245-251.

Older, J. (1986) Low friction arthroplasty of the hip: A 10-12 year follow up study. *Clin Orthop* **211**, 36-42.

Olsson, E., Goldie, I. and Wykman, A. (1986) Total hip replacement. A comparison between cemented (Charnley) and non-cemented (HP Garches) fixation by clinical assessment and objective gait analysis. *Scan J Rehabil Med* **18** (3), 107-116.

Olsson, S.S., Jernberger, A. and Tryggo, D. (1981) Clinical and radiological long-term results after Charnley-Muller total hip replacement. A 5 to 10 year follow-up study with special reference to aseptic loosening. *Acta Orthop Scand* **52** (5), 531-542.

Onsten, I., Carlsson, A.S., Ohlin, A., Nilsson, J.A. (1994) Migration of acetabular components, inserted with and without cement, in one-stage bilateral hip arthroplasty. A controlled, randomized study using roentgenstereophoto grammetric analysis. *J Bone Joint Surg* **76A** (2), 185-194.

Pellicci, P.M. and Haas, S.B. (1990) Disassembly of a modular femoral component during closed reduction of the dislocated femoral component: A case report. *J Bone Joint Surg* **72A**, 619-620.

Russotti, R.P., Coventry, M.B. and Stauffer, R.N. (1988) Cemented total hip arthroplasty with contemporary techniques: A minimum five year follow up study. *Clin Orthop* **235**, 141-147.

Salvati, E.A., Ranawat, C.S., Wilson, P.D. Jr, McCoy, T.H. (1989) A long-term study of Charnley total hip replacements. *Arch Putti Chir Organi Mov* **37** (1), 37-48.

Salvati, E.A., Wilson, P.D. Jr, Jolley, M.N., Vakili, F., Aglietti, P., Brown, G.C. (1981) A ten-year follow-up study of our first one hundred consecutive Charnley total hip replacements. *J Bone Joint Surg* **63A** (5), 753-767.

Schulte, K., Callaghan, Kelley, S., *et al.* (1993) A minimum twenty year outcome of Charnley total hip arthroplasty. *J Bone Joint Surg* **75**, 961-75.

Solomon, M.l., Dall, D.M., Learmonth, I.D., Davenport, J.M. (1992) Survivorship of cemented total hip arthroplasty in patients 50 years of age or younger. *J Arthroplasty,* **7** (Suppl), 347 352.

Stauffer, R.N. (1982) Ten year followup study of total hip replacement. *J Bone Joint Surg* **64A**, 983-990.

Stringa, G., Aglietti, P., Di Muria, G.V., Marcucci, M., Buzzi, R., Lucattelli, G. (1989) Total hip prosthesis according to Charnley. Review of our cases. *Arch Putti Chir Organi Mov* **37** (1), 9-35.

Stringa, G., Di Muria, G.V., Pitto, R.P., Marcucci, M. (1992) Charnley total hip prostheses: clinical and radiographic results after 10 to 18 years. *Chir Organi* **77** (4), 433-435.

Sudmann, E., Havelin, L.l., Lunde, O.D., Rait, M. (1983) The Charnley versus the Christiansen total hip arthroplasty. A comparative clinical study. *Acta Orthop Scand* **54** (4), 545-552.

Sullivan, P.M., MacKenzie, J.R., Callaghan, J.J., Johnston, R.C. (1994) Total hip arthroplasty with cement in patients who are less than fifty years old. A sixteen to twenty-two year follow up study. *J Bone Joint Surg* **76A** (6), 863-869.

Terayama, K. (1986) Experience with Charnley low-friction arthroplasty in Japan. *Clin Orthop* **211**, 79-84.

Wejkner, B., Stenport, J. (1988) Charnley total hip arthroplasty. A ten to 14-year follow up study. *Clin Orthop* **231**, 113-119.

Wejkner, B., Stenport, J. and Wiege M (1988) Ten-year results of the Charnley hip in arthrosis. *Acta Orthop Scand* **59** (3), 263-265.

Wejkner, B. and Wiege, M. (1987) Correlation between radiologic and clinical findings in Charnley total hip replacement. A 10-year follow-up study. *Acta Radiol* **28** (5), 607-613.

Welch, R.B., McGann, W.A. and Picetti, G.D. (1988) Charnley low-friction arthroplasty. A fifteen to seventeen-year follow-up study. *Orthop Clin North Am* **19** (3), 551-555.

Wroblewski, M.B. (1986) 15-21-year results of the Charnley low-friction arthroplasty. *Clin Orthop* **211**, 30-35.

Wroblewski, M.B. and Siney, P.D. (1992) Charnley low-friction arthroplasty in the young patient. *Clin Orthop* **285**, 45-47.

Wroblewski, M.B., Taylor, G.W. and Siney, P. (1992) Charnley low-friction arthroplasty: 19 to 25-year results. *Orthopaedics* **15** (4), 421-424.

Wykman, A., Olsson, E., Axdorph, G., Goldie, I. (1991) Total hip arthroplasty. A comparison between cemented and press-fit noncemented fixation. *J Arthroplasty* **6** (1), 19-29.

Wykman, A., Sanjay, B.K., Soderlund, V., Goldie, I. (1988) Influence of acetabular component position in non-cemented total hip arthroplasty. A radiologic and clinical study. *Acta Radiol* **29** (6), 746-748.

Wykman, A., Selvik, G. and Goldie, I. (1988) Subsidence of the femoral component in the noncemented total hip. A roentgenstereophotogrammetric analysis. *Acta Orthop Scand* **59** (6), 635 637.

3

Evaluation of the Success of Non-Cemented Porous and HA Coated Metal-UHMWPE Total Hip Implant Systems

Sheila Rao
Adam Leckey

INTRODUCTION

Thousands of total hip arthroplasties are performed every year in the United States alone. Medical technology is continually changing to meet the demands for successful, efficient surgical techniques. However, implant design and mechanical behavior also have to be reliable in order to produce a working system. Consequently, materials technology is continually developing new materials to improve hip implant systems. Many materials systems have been tried in the past twenty years. Metal implants were first used for hemiarthroplasties, and in the 1960s Charnley introduced the metal-plastic cemented implant (Charnley, 1972). The concept of the Charnley implant was a 'frictionless implant' (Chapter 2). The implant is composed of a metal femoral component and an acetabular component made of ultra-high molecular weight polyethylene (UHMWPE). Both components are cemented into place using polymethylmethacrylate (PMMA) cement. Since the Charnley implant, numerous other systems have been developed and are currently in use. Non-cemented prostheses represent one such system, in which some of the potential disadvantages of bone cement are eliminated. The purpose of this chapter is to analyze the success rate of uncemented, metal on UHMWPE total hip arthroplasty. The specific types of uncemented total hip arthroplasty (THA) that are analyzed are hydroxyapatite (HA) and porous coated metal-UHMWPE hip implants. The data were collected from published clinical studies of uncemented arthroplasties.

GOALS

The initial goal of this chapter was to compare the success of uncemented, porous metal coated, with HA coated total hip arthroplasties. However, it became clear that the cases that dealt with uncemented implants did not always specify whether the implant was coated with HA or a porous metal, and in many cases the implants had both. Consequently, this chapter will analyze uncemented implants in general in order that their clinical performance can be compared with cemented implants (Chapter 16).

A typical femoral component with areas of beaded metal coating to provide uncemented fixation is shown in Fig. 3.1.

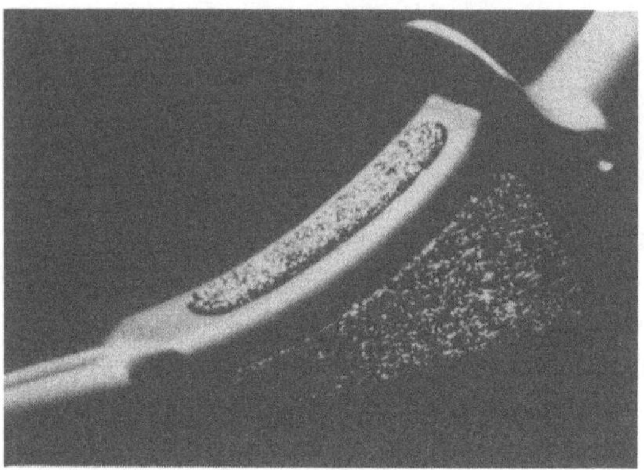

Fig. 3.1 Femoral stem with porous coating (Zimmer Corp.)

METHODS

Published data were used to determine the success or failure of these implant systems. To characterize the data in terms of success or failure, criteria for success were established and are shown in Table 3.1. The cases from each study were evaluated according to these criteria. The numbers of successes and failures were then tabulated and compared with the total number of cases analyzed, to establish a curve which represents the success with time of these implants. Comparisons were made to determine the effect of age, sex, and weight of the patient with respect to success of the implants.

The basis for determining the success of the implant with regard to pain was subjective. In several of the studies that were analyzed, the Harris 100 point scale was used to evaluate the postoperative results. In this analysis a Harris score of 80 or above was considered a success (see Table 2.2). Function was also largely subjective, since it relates to the improvement or

Table 3.1 Factors for success/failure of uncemented THAs

Factor	*Success*	*Failure*
pain	slight to no discomfort due to implant	excessive discomfort-based on Harris score
function	based on improvement from preoperative condition	severe restriction in motion based on Harris score
infection	none	implant-related
fracture	slight rotation no revision required	component loosening fracture/dislocation requiring revision
wear	none	loosening/infection

otherwise of movement before and after surgery. Improvement was considered to be a success, whereas severe limitations, such as being bedridden or unable to perform everyday household activities, indicated a failure.

In terms of infection, cases in which the patient suffered due to an implant-related infection such as those secondary to UHMWPE debris or loss of beads from a porous coating, were considered failures. Patients who suffered from intraoperative infections, or infections due to other circumstances, were not included. Those with infection prior to surgery were also excluded.

Fracture, dislocation and loosening of the implant were all considered failures if they required the patient to undergo revision surgery. In somc cases, the implant loosened due to bone resorption, this was also considered a failure since the implant should be fixed to the bone in such a way that resorption does not occur. Slight rotations of the femoral or acetabular component were acceptable as long as revision surgery was not required. Wear was considered, since wear of the polyethylene cup can cause infection and loosening of the components. If revision surgery was needed due to the wear, or severe infection resulted from the debris, the implant was considered a failure.

The analysis in this chapter did not consider surgical errors and problems in deciding whether an implant was a success or failure.

RESULTS

These results represent a total of 6,117 cases of non-cemented THAs which were followed for up to ten years. The raw data appear in Table 3.2. Figure 3.2 shows the success rate year-by-year based on these data and it shows a marked decline after six years. It seems to suggest an improvement between 9 and 10 years but this is unlikely to be real. Some publications described failures in the 10 year follow-up period without giving the specific time at which they occurred. However, the trend is still clear; as time progresses the implants become less reliable.

DISCUSSION OF LITERATURE

The advantages of non-cemented systems are usually agreed to be their relatively few complications and the ease of performing revision surgery if needed. Originally, non-cemented implants were developed to avoid the loosening that can occur with cemented implants. Non-cemented implants function either by forming a mechanical bond with the host bone, by allowing bony ingrowth into the porous coating of the implant or by direct

Table 3.2 Success of non-cemented THA

Year	*Total Cases Analyzed*	*# Successes*	*% Success*
1	241	237	98.34
2	2678	2539	94.8
3	1088	979	89.98
4	2214	2120	95.75
5	2893	2665	92.12
6	1544	1418	91.84
7	241	177	73.44
8	241	137	56.85
9	357	165	46.22
10	400	256	64

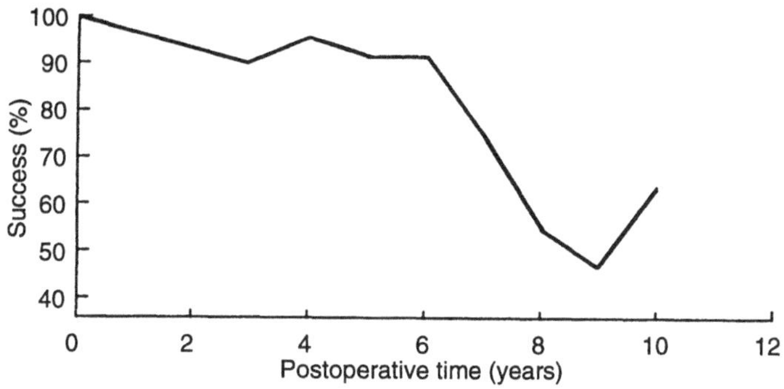

Fig. 3.2 Time dependence success of non-cemented THA.

chemical bonding to a bioactive coating such as HA. Non-cemented implants are proposed as better suited for younger and more active patients, with the expectation that they will last longer without loosening.

However, the cemented implant actually has better long-term results than the non-cemented despite the risk of infection or component loosening (Chapter 16).

The success rates of non-cemented prostheses have been graphically demonstrated in several published cases. Figures 3.3 and 3.4 compare these data, which represent survival rates for over 500 porous coated non-cemented hip implants (Owen *et al.,* 1994).

Young males encounter more problems with aseptic loosening, according to various authors, perhaps because of stress on the system due to higher expectations of performance. Weight is also a determinant, the weight of the individual controlling the load that the implant has to carry.

COMPARISON OF PUBLISHED RESULTS

The conclusions from our analysis do not agree with many published comments on uncemented THAs. Table 3.2 and Fig. 3.2 show that after 6 years the implant is less than 75% successful. This suggests that this type of implant does not provide a solution for anyone who expects more than 6 years of active life.

In this analysis 55.2% were women and 44.8% were men. The average age was 57. However, the age range was very wide and studies rarely presented failures in terms of age or sex.

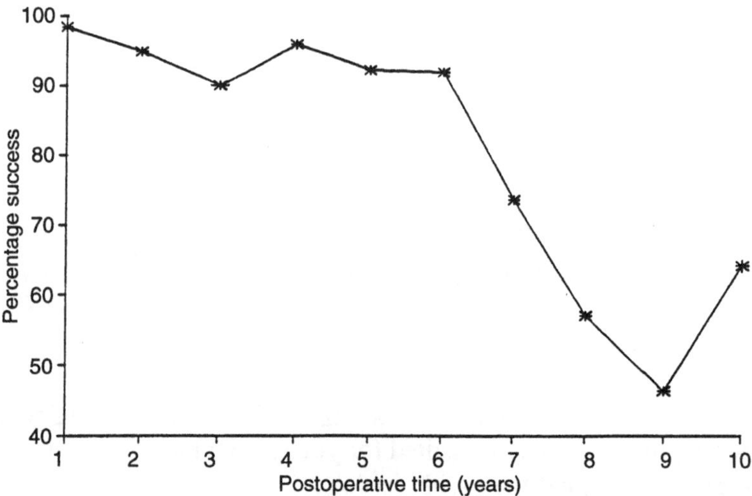

Fig. 3.3 Time dependent success of uncemented THA with mechanical and bioactive bonding (HA coating).

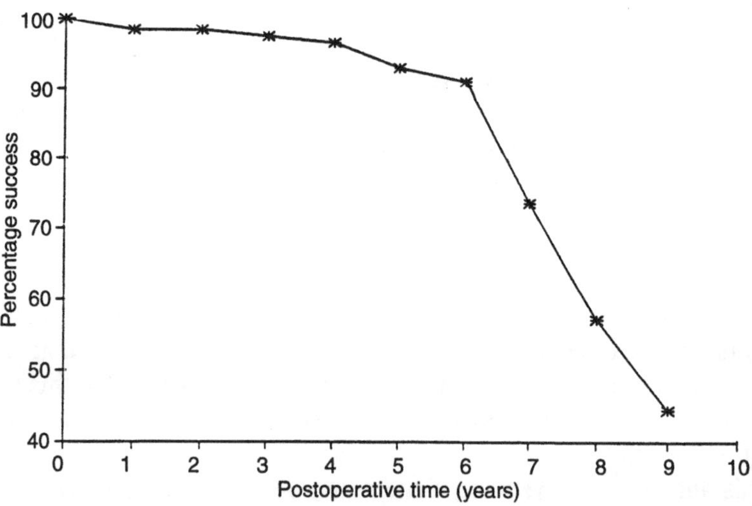

Fig 3.4 Time dependent success of uncemented THA, mechanical bonding only.

It was not possible to analyze the data to compare the survival of the femoral versus acetabular components, since most studies did not distinguish between femoral and acetabular failures, although Morscher (1983) concluding that non-cemented implantation would eventually replace cemented, at least in the acetabular component. However, at this time the data do not support this conclusion.

These data may be discouraging because the criteria for success are strict; however the determination of whether an implant was a failure or a success is the same as is applied to cemented prostheses in other chapters. The combined data reported herein are consistent with the recent report of Havelin *et al.* (1994) comparing cemented vs uncemented THA with 15,335 prostheses involved. They found at five years a revision rate for non-cemented THAs twice that for cemented THA. The difference was more pronounced for patients <60 yrs., especially men. The revision rate of femoral components was three times that of acetabular components.

It is very difficult to determine precise numbers and causes of failure in the studies used for the analysis in this chapter. It was often difficult to believe that the authors were entirely objective when comparing their results with others.

In order to obtain better data on non-cemented metal on plastic total hip arthroplasty, more long-term studies need to be reported. These studies should be conducted in such a manner that the failures are defined in terms of the cause of failure, the time to failure, and age, sex, and weight of the patient.

REFERENCES (Those not cited in text also provided data for Table 3.2)

Charnley, J. (1972) The long-term results of low-friction arthroplasty of the hip performed as a primary intervention. *J. Bone Joint Surg* **54B** (1), 61-76.

Engh, C.A. (1983) Hip arthroplasty with a Moore prosthesis with porous coating, a five-year study. *Clin. Orthop.* **176**, 52.

Engh, C.A. and Bobyn, J.D. (1988) The influence of stem size and extent of porous coating on femoral bone resorption after primary cementless hip arthroplasty. *Clin. Orthop.* **231**, 7.

Geesink, R.G.T. (1990) Hydroxyapatite-coated total hip prostheses two-year clinical and roentgenographic results of 100 cases. *Clin. Orthop.* **261**, 39.

Havelin, L.L., Espohaug, B., Vollset, S.E., Engesaster, L.B. (1994) Early failures among 14,009 cemented and 1,326 uncemented prostheses for primary coxarthrosis. *Acta Orthop Scand* **65** (1), 1-6.

Hernandez, J.R., Keating, E.M., Faris, P.M. *et al.* (1994) Polyethylene wear in uncemented acetabular components. *J. Bone Joint Surg.* **76-B**, 263-66.

Judet, R., Siguier, M., Brumpt, B. and Judet, T. (1978) A noncemented total hip prosthesis. *Clin. Orthop.* **137**, 76.

Keggi, K.J., Huo, M.H. and Zatorski, L.E. (1993) Anterior approach to total hip replacement: surgical techniques and clinical results of our first one thousand cases using non-cemented prostheses. *Yale J of Biology and Medicine.* **66**, 243-56.

Malchau, H., Herberts, P. and Ahnfelt, L. (1993). Prognosis of total hip replacement in Sweden - follow-up of 92,675 operations performed 1978-1990. *Acta Orthop Scand.* **64** (5), 497-506.

Morscher, E.W. (1983) Cementless total hip arthroplasty. *Clin. Orthop.* **181**, 76.

Owen, T.D., Moran, C.G., Smith, S.R., Pinder, I.M. (1994) Results of uncemented porous coated anatomic total hip replacement. *J Bone Joint Surg.* **76-B**, 258.

Ring, P.A. (1978) Five to fourteen year interim results of uncemented total hip arthroplasty. *Clin. Orthop.* **137**, 8.

Tullos, H.S., McCaskill, B.L., Dickey, R., Davidson, J. (1984) Total hip arthroplasty with a low-modulus porous-coated femoral component. *J. Bone Joint Surg.* **66-A**, 888.

4

Alumina-Alumina and Alumina-Polyethylene Total Hip Prostheses

Julie A. Miller
James D. Talton
Sameer Bhatia

INTRODUCTION

By the late 1960s, the Charnley hip implant system with a metallic femoral stem and an ultrahigh molecular weight polyethylene (UHMWPE) acetabular component was the primary design for successful total hip arthroplasty (THA). Many of today's concerns regarding the Charnley system were recognized as early as 1970. Polyethylene wear and its consequences, failure of polymethylmethacrylate (PMMA) bone cement, and long term implant durability for younger patients are all issues that still affect hip implant research and development. These problems led many investigators to search for improved bearing surfaces and means of cementless fixation.

In 1972, Boutin reported on the use of dense alumina ceramic as a bearing surface for total hip arthroplasty and he performed his first procedure with this new device in 1970 (Boutin, 1972). By 1974, several different research teams were investigating the potential of alumina-alumina hip systems. It was estimated that over half a million hip protheses with alumina components were implanted by 1992 (Hulbert, 1993). By 1994, 1.6 million Biolox® ball heads had been used clinically (Willmann, 1995). The components are shown in Fig. 4.1. But, as reviewed by Clarke (1992), ceramic components for hip implants have not been extensively used clinically in the United States. Only ceramic femoral heads, articulating with UHMWPE sockets have been approved for clinical testing. Most research in the field has been conducted in Europe.

®Cerasiv Gmbh. Plochingen, Germany.

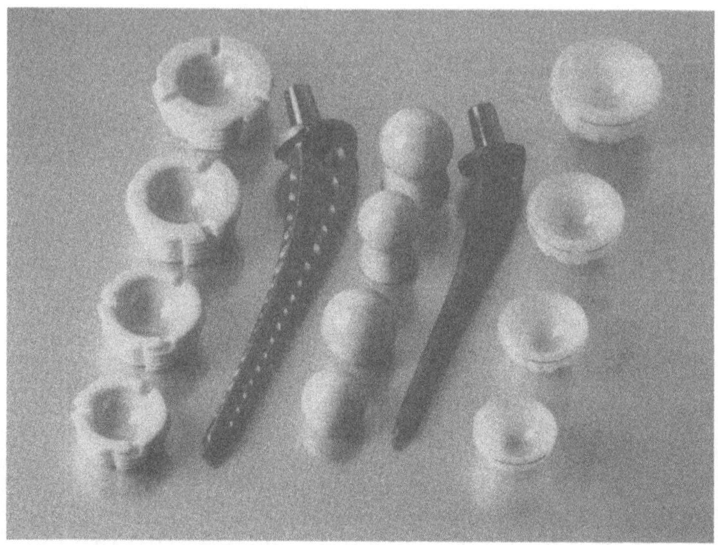

Fig. 4.1 One of the first THA systems using ceramic heads and cups (after H. Mittelmeier). Photo courtesy G. Willmann, Cerasiv Gmbh.

Alumina is a useful material for orthopaedic implants due to its low coefficient of friction, high strength, thermodynamic stability, chemical inertness, and insolubility (see Table 4.1). There are three articulating systems currently in use; metal on UHMWPE, polished alumina on UHMWPE, and polished alumina on alumina. The coefficients of friction for the three systems are 0.16, 0.06-0.10, and 0.03-0.06, respectively. The lower coefficients of friction for the alumina systems help to decrease the problem of wear at the articulating surface by almost 100μm per year over metal systems (Dorlot *et al.*, 1989). Furthermore, high purity (>99%) alumina is a polycrystalline solid in its highest oxidative state (Mittelmeier and Heisel, 1992). This distinguishes ceramics from metallic and polymeric biomaterials, i.e. it is not susceptible to passivation or oxidative degradation *in vivo.* Medical grade alumina must have grain sizes of less than 7 μm, a low average surface roughness and no inclusions (ISO Standards 130/D13 6474). Early problems associated with implants made of alumina were primarily due to tensile failure of the femoral heads (hoop stresses) and wear. These problems have been overcome by improvements in the design of implants and materials processing which leads to a uniform small grain size (Hulbert, 1993).

Table 4.1 Properties of alumina ceramics used in surgical implants

Property	*Unit*	Al_2O_3
Purity	%	>99.7
MgO	%	<.3
Density	g/cm^3	3.98
Grain Size (average)	μm	3.6
Bending Strength	MPa	595
Compressive Strength	MPa	4250
Young's Modulus	GPa	400
Hardness	HV	2400
Fracture Toughness K_{Ic}	$MN/m^{3/2}$	5

Early comparisons of ceramic versus metal alloy femoral heads were based on technical aspects as well as wear performance. The alumina/UHMWPE system has been shown to have superior wear properties over cobalt-chromium/UHMWPE system (Hulbert, 1993). With the proven strength of well manufactured alumina balls there appears to be no mechanical or structural disadvantage when combined with optimally designed neck tapers (Willmann, 1995). A reduced incidence of loosening of both components is expected after long-term replacement because of the low frictional and wear-resistant properties of alumina as well as the shock-absorbing characteristics of UHMWPE.

The objectives behind alumina-alumina cementless hip prosthesis development were to provide a self-locking system with a screwed ceramic acetabular component and a design with interchangeable head and femoral neck sizes (Mittelmeier and Heisel, 1992). The basic design of this class of implants consists of a ceramic head with a standard diameter of 32 mm secured to the metallic femoral stem via a self-locking cone with a 2° 50′ angle and a taper linearity between 0.5 and 0.75 μm. Together with the acetabular component, the ceramic head has a radius tolerance of 7-10 μm and a sphericity deviation of ± 1 μm (Boutin *et al.*, 1988). There are two basic designs for the alumina acetabular component. The solid alumina cup has 3 equatorial pegs and an outer cup each with 1 mm surface grooves for screw fixation. There is also a metal backed alumina cup which is seated in a titanium ring for fixation (Boutin *et al.*, 1988). These alumina head and acetabular components are used with both cobalt-chrome and titanium femoral stems.

Non-cemented alumina components can be well fixed and perform well in the presence of adequate, healthy bone stock, but that does not always occur. However, the use of uncemented devices with a large elastic modulus may result in migration of the acetabular socket and early postoperative failure. Certain groups have advocated the use of cemented sockets made of alumina, so that the modulus mismatch between the bone and alumina can be reduced. The alumina-alumina cemented sockets have the same basic design as the non-cemented prostheses. The benefits of cemented fixation include immediate stability, reduction in blood loss, and excellent pain relief. On the other hand, if the probability of revision is significant, cementless insertion should be considered (Parhofer and Monch, 1984).

UHMWPE acetabular components of sizes ranging from 42 to 52 mm are used with 28 mm alumina heads placed on cobalt-chromium metal stems. UHMWPE acetabular cups contain external grooves that interlock with the cured polymethylmethacrylate cement. In general, cups are rectangular or dove-tailed in profile with a significant bevel to ensure a strong mechanical interlock and prevent subluxation.

BACKGROUND

In order to evaluate the clinical success of the three alumina systems, it is necessary to examine patient demographics, clinical indications for total hip arthroplasty and the general surgical procedure for device implantation. Literature was collected representing clinical data over the past two decades.

There are many different clinical indications for THA, including degenerative disease, trauma and metabolic disease. The majority of primary alumina-alumina hip prostheses have been used to treat four major conditions: primary osteoarthritis, secondary osteoarthritis, rheumatoid arthritis and ankylosing spondylitis, and avascular necrosis.

Minor indications for primary THA are congenital dislocation, fracture and nonunion, and pigmented villonodular synovitis. The major indicator for revision of total hip arthroplasties using alumina-alumina, is aseptic loosening of the cemented hip prostheses. Aseptic loosening of non-cemented devices and multiple revisions are other indicators for additional surgical use of alumina devices. Clinical indications for the alumina-alumina hip arthroplasties are presented in Table 4.2.

Data was obtained for 535 patients who received a total of 575 alumina-alumina non-cemented hip prostheses. Males represented 44.26% of the patients (235) while females composed 55.74%/(296) of the total. The mean age of patients, as published, was 51 years, with a minimum of 21 years and a maximum of 86 years. In at least one of these studies, only patients under 65 years of age were included because of the need for good quality bone stock to achieve cementless fixation. The majority of all arthroplasties were primary in nature (80.69%). Of the 575 total implants, only 111 were used for revisions.

Nine hundred and sixty-four cases of alumina-alumina cemented hips were followed. These studies had nearly balanced ratios of male and female patients with an occasional slight predominance of male patients. Sedel *et al.*, (1990) reported on the use of alumina-alumina devices in patients under 50 years of age, their patients range in age from 18 to 50 years, with a mean of 43 years. Nizard (1992) had a large number of older patients, from 18.7-92.1 years, with a mean of 64.8 years.

For alumina-polyethylene cemented prostheses, 90 implants were reported in 83 patients. Nine males made up 10.84% of the patients while 74 females made up 89.16% of the total. The mean age was 61, with a range of 31 to

Table 4.2 Clinical indications for alumina-alumina THA

Primary Alumina-Alumina THA	*number of procedures*	*% of total number*
primary osteoarthritis	225	48.50
secondary osteoarthritis	77	16.59
rheumatoid arthritis/ankylosing spondylitis	42	9.05
avascular necrosis	87	18.75
other	33	7.11
Revision Alumina-Alumina THA		
aseptic loosening: cemented	101	91
aseptic loosening: non-cemented	2	1.8
multiple revisions	8	7.2

70 years. The level of daily activity was comparable between the two groups, most female patients were housewives and most men were retired.

SURGICAL TECHNIQUE

The general surgical procedure for implantation of the alumina systems consists of a standard approach to the affected hip joint, usually posterior or transtrochanteric, followed by osteotomy of the femoral neck. Articular cartilage and bone are resected from the acetabulum to obtain a conical shape matching the implant. The cavity in the acetabulum is widened with a rasp corresponding to the size of the selected artificial cup. Threads are cut in the remaining bone and the acetabular component of the implant is screwed into place.

Cemented hip sockets are implanted slightly differently. The acetabular cavity is reamed until it is large enough to accommodate the socket component, cement is placed in the hole and then the socket is firmly placed. Some groups have used a slightly different technique in which the acetabular reaming is done until it is slightly undersize then after application of bone cement, the socket was firmly implanted.

The femoral shaft is prepared by removing the cancellous bone from the intramedullary canal in the proximal femur. This cavity is also enlarged with a rasp to fit the size of the femoral stem. The stem is seated and hammered into place. Finally, the ceramic head is inserted and locked onto the metal stem and the hip joint is reduced to its normal position. Autologous bone grafts are often used, when available, to pack both the acetabulum and the proximal femur prior to implant insertion.

RESULTS

In order to compare results from many different studies and investigators, it is important to examine the methods employed for data collection and evaluation. There are several common methods of evaluating patients, including physical examination, radiographic analysis, subjective evaluation, and scoring methods such as the Harris hip score scale or the Charnley grading system (see Chapter 2). Radiography is used to study the implant and its interface for signs of loosening and different variables of proper implantation (socket angle, position, lateral coverage, cement thickness). In addition, radiography is used to monitor the sockets for signs of migration, calcar resorption, and stem subsidence. Table 4.3 gives examples of the different evaluation techniques used in the studies.

Comparison of results obtained from different studies is possible only if a standardized definition of success is used by researchers. The definition of success varies among investigators and patients, based on their expectations and criteria for evaluation. However, as stated by Riska (1993): 'To maximize objectivity in evaluating the results of treatment, every case needing a revision operation was classified as a failure'. Based on the collected data, the criterion for failure in this chapter was set as the reported need for revision surgery. It is a consistent method even though it may overlook clinical and radiographic signs of impending failure. This method, known as survivorship analysis, simplifies comparison of data among studies. Data analysis by this method takes into account the loss of patients to follow-up and describes failure rates with respect to time elapsed since surgery.

All clinical studies lose patients to follow-up for many reasons, which can include death. Some reviewers have advocated eliminating these patients from the calculations of the success rates for the study group at the time at which these patients are lost to follow-up. This method is not used by all groups and very few groups report the time when the patients were lost.

Table 4.3 Data collection and evaluation for alumina/alumina non cemented systems

Author	*Evaluation Methods*	*Frequency*	*Min - Max (Mean)*
Groher, 1983	physical exam, patient satisfaction survey	not stated	6 mo. - 5 yr. (26 mo.)
Parhofer and Monch, 1984	physical exam, survey	not stated	(16.6 mo.)
Mahoney and Dimon, 1990	physical exam, survey, radiographs Harris hip scores	1-2, 6, 12 mo. and annually	27 mo. - 66 mo. (51 mo.)
Riska, 1993	physical exam, Charnley scores, radiographs	not stated	1 yr. - 7 yr. (3.6 yr.)
Boehler *et al.*, 1994	physical exam, radiographs	3 mo. and annually	10.1 yr. - 15.7 yr. (11.6 yr.)

Wherever possible, additional calculations were done to factor in the effect of patient attrition and to eliminate elevation of success rates by indirectly considering 'lost patients' as successes.

There are many reasons for failure at different stages in the life of a hip prosthesis. Early failures are often caused by infection, dislocation of the joint, and fracture of the femur or acetabulum. The major cause of later failure is aseptic loosening. Late infection is another concern along with failure due to fracture of the device.

ALUMINA-ALUMINA NON-CEMENTED

Of the 575 alumina-alumina non-cemented implants studied in this survey, 535 of them were followed-up in some manner (Fig. 4.2). Implants were lost

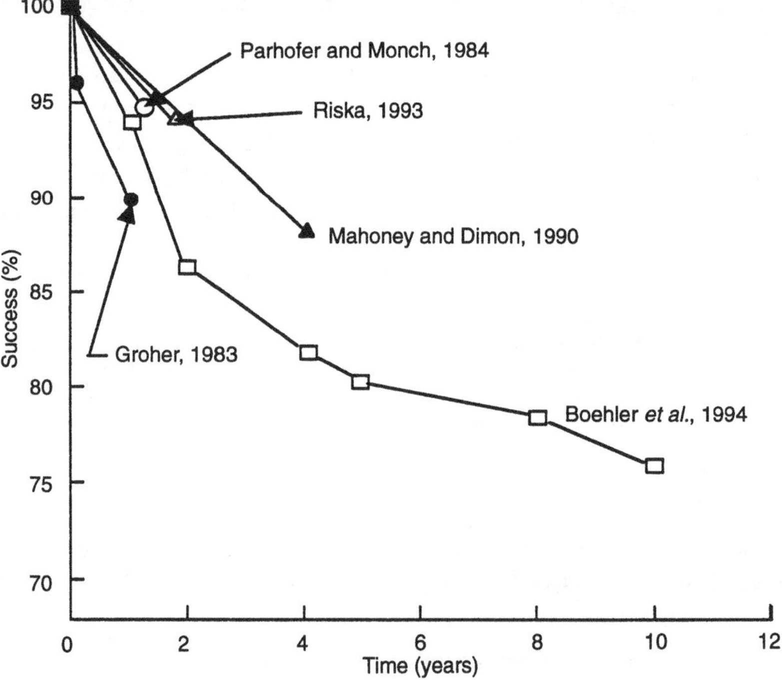

Fig. 4.2 Alumina-alumina non-cemented hip prostheses.

primarily due to patient death from unrelated causes, however many patients simply failed to keep follow-up appointments. In all, implants that could not be followed due to death or loss of patients were excluded from the following evaluation.

In 1983, Groher reported 28 failed implants out of 285 total devices. Each of these failures occurred within the first year of implantation. Ten implants failed in the first 20 days postoperatively due to dislocation (4), acetabular fracture (3), and infection (3). Within one year, an additional 18 implants failed due to aseptic loosening. The success rates for this study are 96.5% and 90.2% respectively at the 20 day and 1 year time periods.

Parhofer (1984) reported a total of 3 revisions out of 55 at an average 16.6 months, a success rate of 94.5%. Mahoney (1990) showed a success rate of 87.9% (3 failures out of 33 patients) at an average of 51 months. Neither of these reports specified reasons for revision surgery. Riska (1993) reported a success rate of 93.8% at 2 years postoperatively. He reported seven failures out of 112 cases, due to aseptic loosening of the acetabular cup (2), late infection (2), pigmented villonodular synovitis (2), and technical failure (1).

The most comprehensive study on alumina-alumina non-cemented prostheses comes from Boehler, in which a ten year follow-up is reported. Out of 50 fully evaluated cases, 38 implants survived for the full ten years (76% success rate). Aseptic loosening and ceramic head fracture were the two reported failure modes. Revision for aseptic loosening of the acetabular component occurred at 1 year, 2 years and 10 years with one failure at each time period. Ceramic head fracture occurred at 1 year (2), 2 years (3), 4 years (2), 5 years (1), and 8 years (1). The overall success rate declined from 94% at one year to 86% at two years. Each decline thereafter was less than or equal to two percent per year (Boehler *et al.*, 1994).

ALUMINA-ALUMINA CEMENTED

Of the 964 cases studied in the survey, the cases were followed for periods ranging from 1 year to 12 years (Fig. 4.3). Patient death and the failure of certain patients to return for follow-up examination contributed to high attrition rates in certain studies. Causes of failure and the time of failure are not reported and make determination of the causes of failure at specific time periods difficult. The reported failures commonly occur due to aseptic

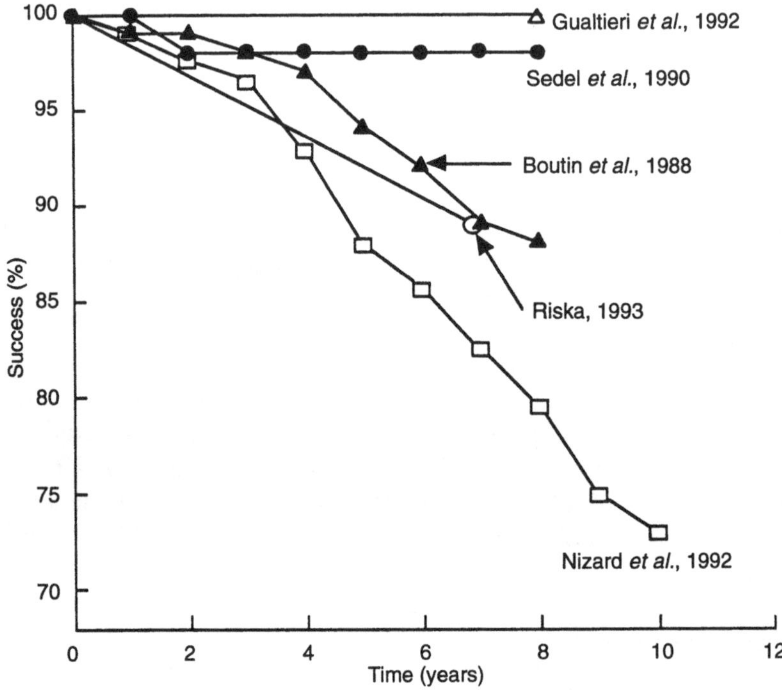

Fig. 4.3 Alumina-alumina cemented hip prostheses.

loosening of the acetabular socket, septic loosening, acetabular socket fracture, femoral head fracture, loosening of the femoral stem and loosening of both components. Certain groups consider failure as a "failure of technique" and not a failure of the material.

Nizard *et al.,* (1992) reported on the survival of cemented alumina-alumina implants in 187 patients followed for up to 10 years. Analysis showed that 82.54% of the implants survived 10 years. Twenty four of these hip implants failed during that time. Fifteen of the twenty four (62.5%) were due to aseptic loosening. Very few cases of femoral component loosening occurred. Specific cases of head fractures due to mismatch between head and socket were also noted at 2 and 3 years. The study also reported on the effect of age on the success rates. Higher success rates were observed in younger patients, under 50 years old.

Sedel *et al.*, (1990) reported the survival of implants in patients under 50. Of 86 hip prostheses implanted in 75 patients, 56 were followed up for periods extending to 12 years. Of these, 54 were followed for more than 2 years and 34 for more than 5 years. Only 1 implant failed due to fracture of the head, 2 years after implantation. Failure was attributed to improper placement of the socket during surgery. Survival was 97.8%.

Boutin *et al.*, (1988) reported on the results of 560 cases of hip prostheses for 1-9 years with a mean of 4.5 years. The survival rates for the study were 88% at that time. Failure rates for 83 patients under 50 years of age were lower than those for older patients in which no failures were reported.

Riska (1993) reported on success rates in 143 patients followed up for 1-12 years with a mean of 6.7 years. Only 16 patients had failures, 12 of them due to aseptic loosening.

Gualteri *et al.*, (1992) reported on the survival rates of implants in 36 cases of ankylosing spondylitis followed up over an average period of 7.8 years. No failures were noted in any of the cases.

Success vs failure rates of the alumina-alumina cemented systems in patients under 50 and those older than 50 are summarized in Fig.4.4. Greater success is seen in the younger patients.

ALUMINA-POLYETHYLENE CEMENTED

Of thc 107 alumina-polyethylene cemented implants studied in this survey, 83 (77.56%) were followed for up to eleven years after surgery (Fig 4.5). Implant or patients were lost due to postoperative infection, loss to follow-up, and patient deaths from unrelated causes.

In the Saito study (1992), only one hip of 57 required revision surgery six years after the initial arthroplasty, as a result of femoral component loosening. One case of dislocation occurred after two weeks but was successfully treated with closed reduction. One case of proximal femoral cracking occurred but was corrected with cerclage wires. One patient suffered a spiral fracture around the femoral component in a traffic accident three years after arthroplasty and was treated conservatively. Four acetabular components in four patients showed X-ray evidence of loosening, but continued to be satisfactory. The average rate of wear for polyethylene was 0.10 mm per year, with a maximum of 2 mm, five years after surgery.

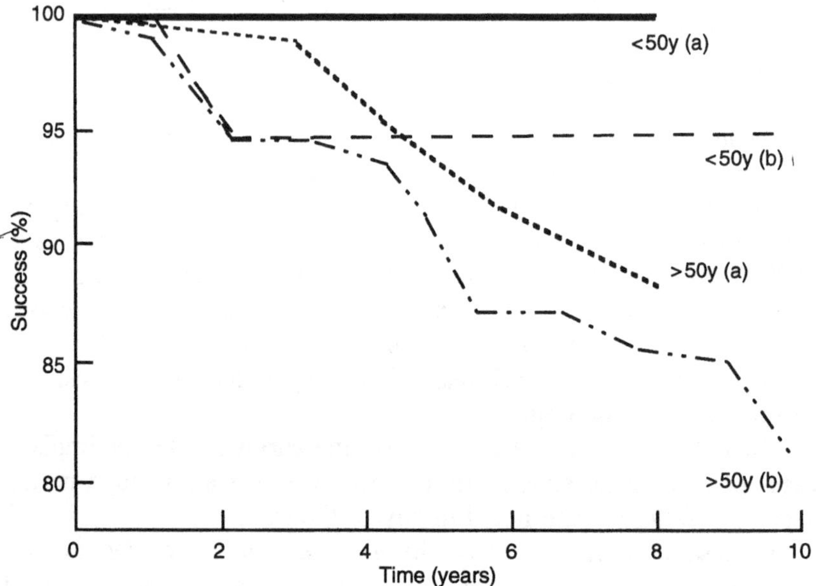

Fig. 4.4 Clinical success of cemented alumina-alumina THA.

a) Boutin *et al.* (1988) — 83 patients under 50 yrs (100% success); 477 patients over 50 yrs (88% success)

b) Nizard *et al.* (1992) — 21 patients under 50 yrs (95% success); 166 patients over 50 yrs (88% success)

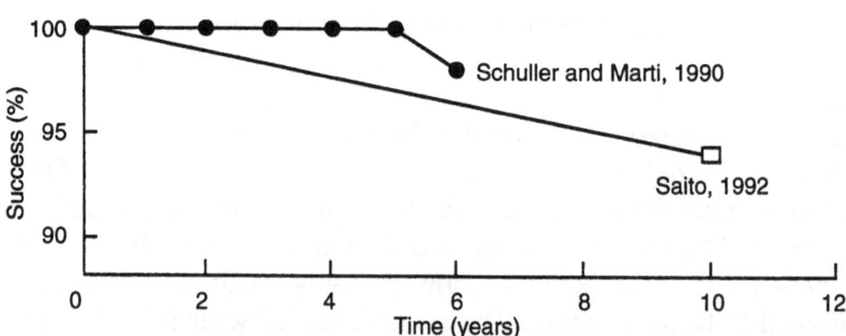

Fig. 4.5 Alumina-polyethylene cemented hip prostheses.

In Schuller (1990) two hips of 33 required revision surgery in the ten year test period, and three showed evidence of radiographic loosening. One revision required replacement of the polyethylene cup and the other of a cracked alumina head. An average of 0.26 mm reduction of the polyethylene diameter was seen compared with 0.96 mm for polyethylene on metal wear. The metal femoral head control group had three revisions after 10 years, and four cases of loosening.

DISCUSSION

In comparing the different alumina based total hip implant systems, two general trends can be seen from Fig. 4.6. First, alumina UHMWPE seems to have the lowest number of failures needing revision surgeries, but statistical analysis of the two poorly detailed reports do not support this.

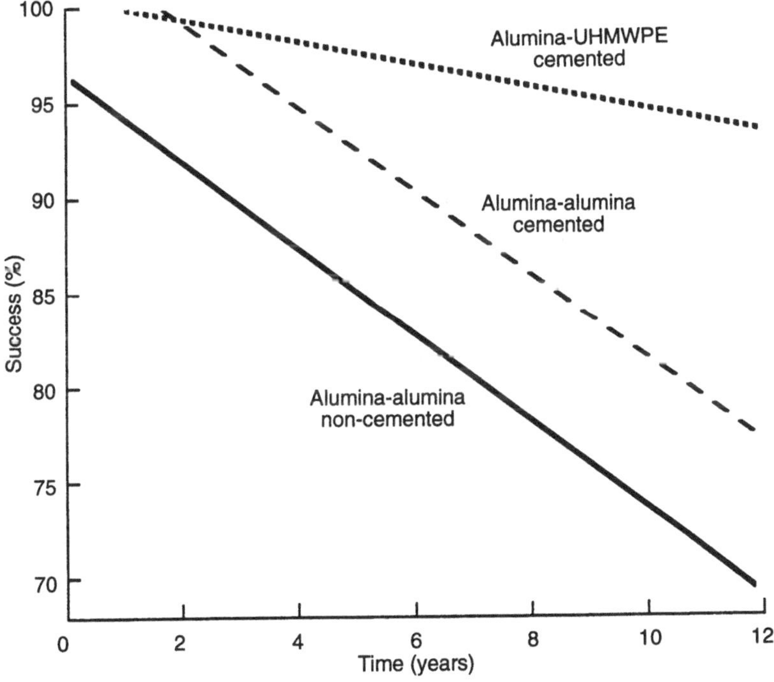

Fig. 4.6 Comparison of alumina hip prostheses.

Willert and Semlitsch, quoted by Mittelmeier, showed that the combination of alumina ceramic balls with polyethylene sockets reduced the wear rate compared with the metal-polyethylene combination, but not as much as alumina with alumina (Mittelmeier, 1992). The results of these tests (88 implants) are not as statistically significant as the alumina-alumina cemented and non-cemented systems (964 and 575 implants respectively).

Second, the cemented alumina-alumina prostheses had a lower incidence of failure than the non-cemented prostheses, about 58 per year up to 12 years, as determined by linear regression. This shows that despite the complications that have been reported with cement curing and aseptic loosening, cement fixation is still very reliable. This is statistically valid, with ten studies reporting from 1 to 12 years of observations.

Correlations between age, sex, preoperative condition, and general activity level could not be derived from these papers. Most reports used a variety of radiographic tests and pain questionnaires, with very few similarities. The general definition of a failure, for comparison, was set at the need for revision surgery, and some studies did not even report this. The activity level and wear seem to be the greatest concern for long-term prostheses.

Overall, of the alumina-based total hip systems, alumina-alumina cemented prostheses are rated the best (20% failures after 10 years), with non-cemented fixation reporting just below this (25 % after 10 years). Alumina-polyethylene prostheses are rated just above those systems, but the number of implants was scarce. Polished alumina femoral heads on a cemented metal shaft with a cemented alumina acetabular cup seemed to perform the best. To quantify these results, a prospective study comparing at least 100 implants of each alumina based system should be done up to 12 years to determine the clinical consequences of fatigue and wear.

REFERENCES

Boehler, M., Knahr, K., Plenk, H., Walter, A., Salzer, M., Schreiber, V., (1994) Long-term results of uncemented alumina acetabular implants. *J Bone Joint Surg,* **76-B**, 53-59.

Boutin, P. (1972) Arthroplastic totale de la hanche par prosthese en alumine fritte. *Rev Chir Orthop*, **58**, 229.

Boutin, P., Christel, P., Dorlet, J.M., *et al.* (1988) The use of dense alumina-alumina ceramic combination in total hip replacement. *J Biomed Mat Res*, **22**, 1203-32.

Clarke, I. (1992) Role of ceramic implants. *Clinical Orthopaedics and Related Research*, **282**, 19-30.

Dorlot, J.M., Christel, P. and Meunier, A. (1989) Wear analysis of retrieved alumina heads and sockets of hip protheses. *J Biomed Mat Res*, **23**, 299-310.

Groher W. (1983) Uncemented total hip replacement. *Can J Surg*, **26** (6), 534-36.

Gualtieri, G., Gualteri, I., Hendriks, M., Gagliardi, S. (1992) Comparison of cemented ceramic and metal-polyethylene coupling hip prostheses in ankylosing spondylitis. *Clinical Orthopaedics and Related Research*, **282**, 81-5.

Hulbert, S. (1993) The use of alumina and zirconia in surgical implants (ed L.L. Hench and J. Wilson) *An Introduction to Bioceramics*, World Scientific Publishing Co. Plc. Ltd., Singapore, pp. 25-40.

Mahoney, O.M. and Dimon, J.H. (1990) Unsatisfactory results with a ceramic total hip prothesis. *J Bone Joint Surg [Am]*, **72-A** (5), 663-71.

Mittelmeier, H. and Heisel, J. (1992) Sixteen-years' experience with ceramic hip protheses. *Clin Orthop*, **282**, 64-72.

Nizard, R., Sedel, L., Christel, P., *et al.* (1992) Ten year survivorship of cemented ceramic-ceramic total hip prosthesis. *Clinical Orthopaedics and Related Research*, **282**, 53-63.

Parhofer R. and Monch W. (1984) Experience with revision arthroplasties for failed cemented total hip replacement using uncemented Lord and PM protheses, in *The Cementless Fixation of Hip Endoprostheses*, (ed E. Morscher) Springer Verlag, Berlin.

Riska, E. (1993) Ceramic endoprothesis in total hip arthroplasty. *Clin Orthop*, **297**, 87-94.

Saito, Masanobu, *et al.* (1992) Efficacy of alumina ceramic heads for cemented total hip arthroplasy. *Clin Orthop*, **283**, 171-77.

Schuller, H.M. and Marti, R.K. (1990) Ten year socket wear in 66 hip arthroplasties. *Acta Orthop Scand*, **61**(3), 240-43.

Sedel, L., Kerboull, L., Christel, P., Meunier, A.,Witvoet, J. (1990) Alumina-on-alumina hip replacement. Results and survivorship in young patients. *J Bone Joint Surg [Br]*, **72-B**(4), 658-63.

Willmann, G. (1995) Production of medical grade alumina. *Br. Ceram Trans*, **94**(1), 38-41.

5

Total Hip Replacement: Metal-on-Metal Systems

Laura Elsberg
Mark Moore

INTRODUCTION

Arthroplasty has been performed since the 1800s, when it was used to correct hip deformity in patients (Jayson, 1971). As the years progressed, material considerations became more important as well as the method of fixation of the components. Total replacement arthroplasty began in the early 1950s, with the work of Charnley and of McKee. McKee began his work with femoral and acetabular components made of stainless steel and cemented into the bone using acrylic cement. He later chose a cobalt-chrome alloy as his material of choice. Advantages of this material choice for the hip prosthesis, according to McKee, include tolerance to body tissues, low wear, and low friction. Loosening is a major concern in his design (McKee and Farrar, 1966). Other types of prostheses include the Stanmore prosthesis, which progressed from all cobalt-chrome (with three pins cemented into the ilium for extra support) to an articulating pair of cobalt-chrome/polyethylene, the Charnley prosthesis, which used ultrahigh molecular weight polyethylene (UHMWPE) for the acetabular component and a metal femoral stem and head; and the Muller self-lubricating hip prosthesis, which was a metal-on-metal system with plastic bearing in the acetabular component for the reduction of friction.

This chapter deals with metal-on-metal systems, with primary emphasis on the McKee-Farrar system. This system is composed of Vitallium®, which is a cobalt-chrome-molybdenum (Co-Cr-Mo) alloy. Other systems existed,

---

®Howmedica Inc. Rutherford, NJ

such as the Mark 5 and Stanmore, but these were not as widely studied. Also cited in this chapter are porous coated metal-on-metal systems, for which there are very few data. This is because, in most studies, UHMWPE is one of the articulating surfaces, either as the acetabular component or the femoral head. UHMWPE components came into widespread use to prevent metal ion release from the metal-metal pairings.

The major failure mechanism of metal-on-metal prostheses is loosening of one or other components. This loosening may be due to cement cracking, cement detachment, or bone resorption. Attempts to control this loosening have included centrifugation instead of hand mixing of the PMMA (polymethylmethacrylate) bone cement (Burke, 1985) and improved cement systems, such as pressurized cement application (Miller, 1985). Other methods of fixation have been explored, such as bioactive and porous coatings for morphological fixation.

Hydroxyapatite (HA) coatings on implants are used to achieve a biological fixation, i.e. a porous coating that bone tissue can grow into and mineralize within. HA coatings are generally applied using a plasma-spray process. HA coatings are considered better for producing earlier and stronger fixation (Oonishi, 1991). However, HA coatings have mainly been in clinical use in the systems that use polyethylene as one of the articulating components.

Porous coatings of metal on titanium or cobalt-chrome alloy hip prostheses were investigated as alternative methods of fixation (Bourne *et al.*, 1994). Improvements in PMMA cement preparation, bone preparation, and improvements in the prothesis design have enhanced the success of the cemented systems in the short and intermediate terms. However, the porous coatings are now being explored for long-term results. Porous coatings allow for tissue ingrowth into the pores, and even mineralization of the bone within these pores, depending on the material. These coatings can either be formed by sintering, plasma-spray deposition, or diffusion. Some systems use one method for the acetabular component and another for the femoral stem, and other systems use a combinations of two of the methods.

While the metal-metal articulating pair lost favor 20 years ago, the system is currently being reexamined, as recent studies have suggested that it may be a successful long-term implant system. Clinical trials have now begun in Germany using metal-metal pairings with porous coatings.

METAL-METAL CEMENTED

The McKee-Farrar system

In 1964, McKee and Farrar selected a cobalt chrome alloy in the shape of a standard Thompson prothesis (one and five-eights inch diameter head) in combination with a cobalt chrome alloy, hemispherical, acetabular cup to fit the head of the Thompson prothesis. The cup had a lip that served the two-fold purpose of reinforcement and prevention of cement overflow. Small studs covered the outer surface of the cup to ensure adequate cement thickness and to allow the cup to have a positive hold when embedded into the cement. Both the cup and the stem were anchored to the bone using methylmethacrylate cement. The components are pictured in Fig. 5.1. The reasons for selection of the cobalt-chrome alloy included low friction resistance, little wear, and inertness of the metal in the body. After further review, the authors modified the Thompson prothesis to have a more slender neck.

Their initial report included quick recovery of the patients from pain, and patients were able to bear partial loads within a week of operation. The results from 47 of the first 50 cases were labeled as 'excellent' or 'good'. There is a more than 90% success rate for the first four years, but they advocate further testing.

To meet this need for further testing, several authors have provided detailed postoperative studies on McKee-Farrar type protheses. They provide various ways of evaluating what could be considered 'success'.

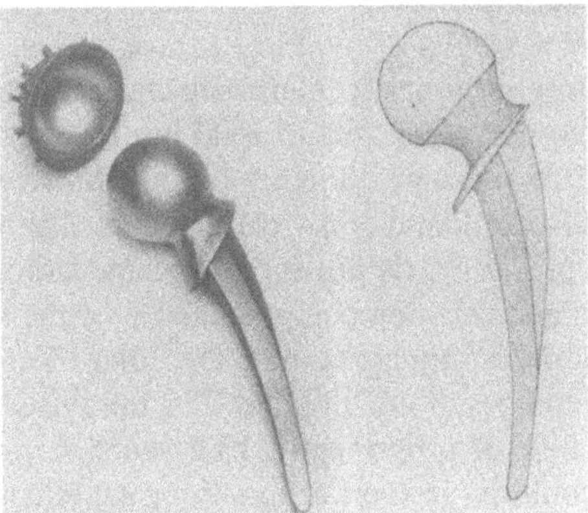

Fig. 5.1 The basic McKee Farrar hip prothesis.

Review of McKee-Farrar published data

One important study was presented by McKee and Chen in 1973. In their study, they detail cases of the various series of the McKee-Farrar prothesis. Of particular interest to this review was the fourth series of the total hip replacement which used a modified Thompson femoral component, and purely cement fixation (1966 and after). The study found that 75 out of the first 100 cases had good or excellent initial results before revision, 86 out of the second 100 and 90 out of the third 100. Based on the evaluation of the three hundred cases, the authors discovered that the main determinant of success with the implants was firm fixation. To address this problem, the system was modified and screws to aid in fixation were reintroduced. The initial results of the first 100 cases of this slightly modified method showed 97 with good or excellent results.

In 1982, Tillberg presented a prospective study of 327 total hip arthroplasties using the McKee-Farrar implant. The median age of the patients at the time of the operation was sixty seven. The patients were followed up each year for a period of up to eight years. Results of the study included information on pain, mobility, and walking capacity graded accorded to Charnley (1972) (see Table 2.1) for preoperation, one year after operation, and seven years after operation. It was found that 90% of patients experienced total freedom from pain 3 months after operation, which lasted for 2 years. Also, it was found that the percentage of good results (Charnley grades five and six) seemed to decrease two years after the operation. The level of good results was approximately 50% at seven years. An important conclusion was that total hip arthroplasty is successful for older patients, but based on the longer term results of total hip arthroplasties, younger patients should first consider hip arthrodesis, intertrochanteric osteotomy, or Smith-Petersen arthroplasty to delay the time of total hip replacement.

August *et al.* in 1986 presented a study on 808 McKee-Farrar total hip arthroplasties done at Norfolk and Norwich hospitals. The evaluation of the patients included use of the Harris hip score system, radiographic analysis, and survivorship. Pain and function comprised 91% of the total score, while movement and lack of deformity constituted the other 9%. After subtracting untraceables, deaths, and revisions, this left 230 unrevised with Harris hip scores, at an average time of follow-up of 13.9 years. A major result of the study was that 71% could be in the fair, good, or excellent range.

The primary purpose of the radiographic results was to see if there was loosening of either the acetabular component or hip stem. It was found that 50% of the femoral components had stem loosening, whereas 51.1% of the acetabular components were loose.

The results of the survivorship curve are presented in Fig. 5.2. Data were available on 657 hips (81.3%). it appears from this survivorship curve that survivorship *in situ* was 84.3% at 14 years, but was around 20% at 20 years. Loosening appeared to be the main cause of failure.

A twenty year study was done on total hip arthroplasties in Greece between November 1966 and July 1974. Zaoussis and Patikas (1989) THAs including 115 McKee hips, 16 low friction arthoplasties, 8 Harris type, 2 Muller, and 2 Howse type. The average age of the patient was 53.4 years with a range of 19-85. In the first ten McKee cases, screws were used in addition to cement. After removal of cases with insufficient data, deaths, revisions, and loosening, 43 *in situ* THAs were available for study. These hips survived a minimum of twelve years.

The surviving hips were clinically analyzed for the presence and of pain, range of passive flexion, muscle power of the operated leg, walking, and function. Long term results included 53% of patients who had

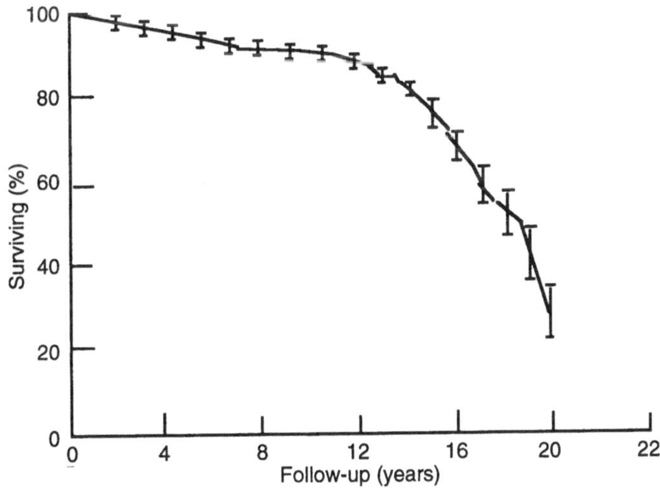

Fig. 5.2 Survival Curve of August *et al.*, (1986).

a complete lack of pain, and there was preservation of useful flexion in 79%. The authors believed, however, that the number of surviving cases (30%) was too small, and that many failures were attributable to poor operating conditions.

Also in 1986, five-year radiographic and clinical studies on 84 McKee-Farrar and 54 Charnley protheses were presented in two papers by Djerf and Wahlstrom (1986). The mean age of the patients undergoing the total hip arthroplasty was 66.4 years ± 8.4 years. Clinical tests included flexion, leg length, walking ability, walking speed, walking aid, pain, and patient opinion. Correlations were made with body weight, age, and the surgeons. Results from the clinical tests showed that more than 90% of the McKee-Farrar patients were free from pain at rest, while walking, and after walking. Also, maximum flexion increased during the first year and did not deteriorate, and walking speed improved throughout the five year followup. However, after a first year of improvement, walking speed decreased. A major conclusion from the clinical studies was that there appeared to be no difference between metal-on-metal and to metal-on-polyethylene after five years.

In the radiographic study, X-ray films in anteroposterior and lateral projections were used postoperatively and every year for five years. Measured variables included valgus/varus inclination, position of the stem tip, lateral opening, anteversion, cover, and thickness of the femoral cortex. Thirty-nine percent of the McKee-Farrar and 24% of the Charnley prostheses showed no loosening at five years. However, there was no significant difference between the Charnley and McKee-Farrar in terms of lateral opening of the cup, and varus or valgus positioning of the hip.

In 1987, Visuri detailed a study of 511 McKee-Farrar total hip replacements done at the Orthopaedic Hospital of the Invalid Foundation in Helsinki, Finland from 1967 to 1973. Three types of implants were evaluated including one with a one and five-sixths inch bearing diameter, and two with a one and three-eighths inch bearing diameter. Of the two systems with smaller bearing diameters, one was paired with a stem with a short neck and the other was paired with a stem with a long neck. Reinforcement screws were required in 121 cases.

The age of the patients ranged from 22 to 79 years with an average age of 59.4 years. Survivorship tables were constructed by the Armitage method modified by Dobbs (1980). Risks factors for aseptic loosening investigated included sex, type of prothesis, postoperative time etc.

The survivorship curves are presented in Figs. 5.3 and 5.4. It is apparent that the small bearing diameter with the long neck had the best survival curve. Risk factors that could be considered significant included the type of implant, right hip prothesis (dominant walking leg in most people), lengthening of the leg in operation. Weight, age, and physical activity were not significant.

Revision was a major topic of a study done by Pepten *et al.* in 1989. In the study, they compare the survival rates of implants after revision to overall survival curves of various implant types. Fig. 5.2 shows the survivorship of the McKee-Farrar prothesis done by August *et al.* (1986).

In 1991, Jantsch *et al.* presented a long term study on McKee-Farrar using the Mayo Clinic hip score. A total of 100 total hip protheses that remained *in situ* were followed up clinically and/or radiographically. There were also 36 cases of revision arthroplasties followed up. The mean age was 70 years at the time of the last measurement, with a mean follow-up time of those without revisions of 14 years. The results of the clinical tests show that for 78% of the patients without revision, results were satisfactory. Radiographically, 66 % of the cups for cases without revision were stable, and 74% of the stems. Combined radiographic and clinical results show that 65% of the implants without revision are satisfactory. The authors felt that the bone-cement interface was a weak point.

In 1994, Visuri *et al.,* published a study on the life expectancy of patients with total hip arthroplasty. Patients who received a McKee-Farrar, Brunswick, and Lubinus THA were compared with a control group, matched for age and sex in the same hospital, with other orthopaedic conditions. The study found that there were no differences in mortality between patients with different types of prostheses, nor was there any difference in mortality between the control patients and those with THAs.

STANMORE TOTAL HIP PROSTHESIS

In addition to McKee-Farrar type hip implants other types of metal-on-metal exist, such as Stanmore and Mark five.

In 1980, Dobbs presented a study on total hip replacement using the Stanmore metal-on-metal hip components and Stanmore metal-on-plastic hip components. Using the method described by Armitage (1971) survivorship

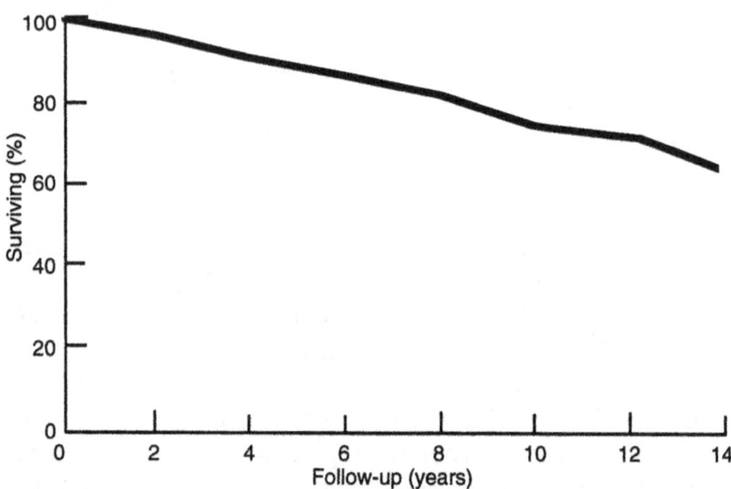

Fig. 5.3 Composite survivor curve of Visuri (1987).

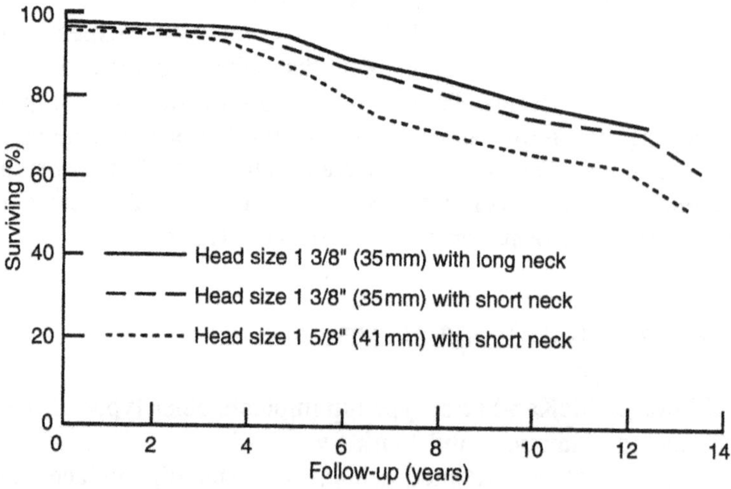

Fig. 5.4 Individual survivor curves of the systems in Visuri (1987).

tables were constructed for both cases. Also, graphs of the survivorship curves and overall probability of failure were presented for both cases. Specifically for metal-on-metal, the overall survival rate was 53% after 11 years. The annual removal rate was 5.5% per year. The highest percentage of removed metal-on-metal implants was because of loosening.

Two important assumptions of the Armitage calculations for life time were that withdrawals (deaths, etc.) had the same removal probability as non-withdrawals, and the annual probability of removal must remain constant over time. The latter was because patients entered the study at different times (improved operative technique, different doctors, etc.).

POROUS COATINGS TO REPLACE CEMENT

Methods of application

As previously stated, the porous coating can be applied by one of three techniques.

The sintering technique results in a porous coating composed of layers of spherical beads (either CoCr or Ti), differing only in bead size, bonded to a cobalt-chrome or titanium substrate. The bond strength of the porous layer is controlled by the temperature and time of heating during sintering. Porosity, necessary for bone ingrowth, is regulated by the original bead size.

Diffusion bonding requires that a Ti-fiber mesh be prefabricated and shaped into the desired form. Then, heat is used to attach the mesh to the acetabular component. The heat used is less than that for sintering. Porosity is controlled by the configuration of the wire as well as by the temperature and pressure involved in the process.

The plasma-spray process is like a welding process. A gas is ionized into a plasma spray, and the particles to be deposited on the surface of the substrate are injected into this flame. The flame partially melts the particles, mainly beads, so that they adhere to the implant upon deposition and cooling. Because the implant is not heated by this process, there is no possibility of weakening of the system due to elevated temperature. The size of the pores is controlled by initial particle size and temperature of the plasma spray.

Published data

All studies done on HA coated metal implants were either for the acetabular component, the femoral component, or a metal-UHMWPE system, and no data could be generated for the total system. However, Furlong and Osborn in 1991 report that post mortem specimens from patients with HA ceramic coated, titanium prostheses show that biological fixation is achieved within seven weeks.

Defining success

There were two main approaches to the definition of success. Success in some studies was derived from clinical information, including patient interviews. Methods such as hip scoring and survival times are two examples which are frequently used. However, other studies viewed only the radiographic results of the hip implants. If the hip implant was radiographically loose, the implant was expected to have a higher probability of failure, even though the patient felt no pain and lost no mobility. Some studies use both approaches, but are not presented together.

Patients may define success in terms of the pain they feel or the necessity for revision surgery as well as mobility and time of recuperation. The immobilization time postoperatively is crucial for the most elderly patients, who need to be mobile soon after surgery for other reasons.

Success should be a combination of the clinical and radiographic results. However, the radiographic results are a predictor of future failure, not of current failure. Therefore, for the present evaluation of success, radiographic loosening is excluded. Success for the McKee-Farrar hip replacement is evaluated by using a composite survivorship curve of the studies in the literature review, modified by clinical evaluations (such as hip scoring). Specifically, the magnitude of success at a given time will be the percent surviving with hip scores in the fair, good, and excellent range of the evaluation.

Compiled results

The specific method for evaluating the 'success' of the metal-on-metal implant over a period of 20 years was done using the following steps. The first was to compile and average the survivorship curves of Dobbs (1980), August *et al.* (1986), and Visuri *et al.* (1994). Since the August study extended to 20 years, and the others did not, rates of decline from the August paper were used to extrapolate the other curves. After this step the survivorship curves were multiplied by the fraction of fair and better survival statistics in August *et al.* (1986), Visuri *et al.* (1994), McKee (1982), and Zaoussis *et al.* (1989). Average levels of these results were computed for 0-5 years, 5-10 years, and 10 plus years. Modifications were applied accordingly, and the chart appears as Fig. 5.5.

Schutzer and Harris (1994) cited results of porous coatings on Ti by plasma spraying of HA coatings on porous Ti-alloy implants. The article cites 56 total hip arthroplasties performed on 51 patients, 49 of which were revisions. The patients ranged in age from 26 to 81, with an average age of 51. Of the 49 revision procedures, 30 were because of migration with and without loosening, and 14 implants loosened. Five femoral heads had failed. Overall, they cite only 46% of the 56 implants had congenital dysplasia or dislocation as the original diagnosis. The average Harris score was 86, with a range of 36-100. Kelley (1994) investigated high hip implants. The average Harris hip score for these implants was 78, although again, no distinction was made between excellent, good, fair, and poor scores. While complex hip arthroplasty is more difficult in terms of surgical techniques, the models are adequate and should function as well as the regular THAs. Brinker *et al.*, (1994) cited 66 hip revisions. The implants were divided between Ti and Co-Cr. The group was of mixed race and gender, and scores averaged 87.9 after 68 months. The data were split as follows: 55% above 90 (excellent), 10% between 80 and 90, 5% between 70 and 80, and 11% below 70.

The most successful of all metal THA systems appears to be the McKee-Farrar metal-on-metal total hip replacement system. This system has a better long-term success rate, and several authors recommend that research be focused on making perfectly articulating concentric cobalt-chrome components.

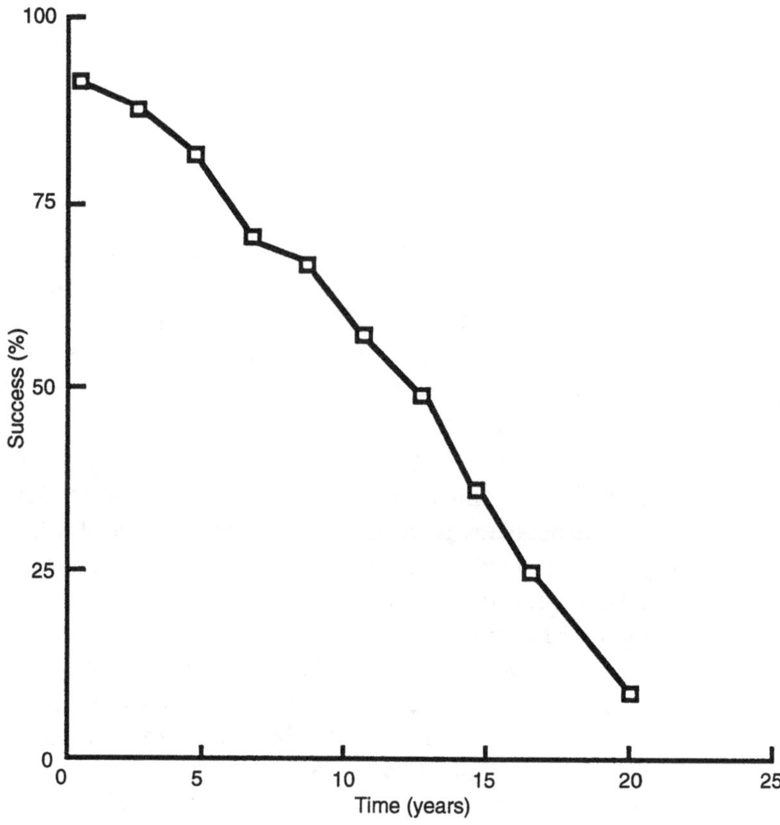

Fig. 5.5 Composite success vs. time for metal/metal THA (success=average survival rate at given time times the average fraction of satisfactory or better arthroplasties at that time).

CONCLUSIONS

There is a great need for a central data base to standardize the results of THA. Each hospital has its own method of summarizing cases and defining success. Many case studies describe either the acetabular component or the femoral component without noting results for the other. Total systems need to be evaluated before recommendations are made.

As stated by Muller in 1992, who developed a total metal-on-metal system, 'Polyethylene is the weak link in THA and ought to be replaced by perfectly concentric metallic sockets and femoral heads of cast Cr/Co/Mo alloys.' Only time will tell whether this recommendation is justified.

REFERENCES

Armitage, J. (1971) *Statistical Methods in Medical Research*, Blackwell, Oxford, pp. 408-14.

August, A.C. *et al.* (1986) The McKee-Farrar hip arthroplasty, *J Bone Joint Surg*, **68-B** (4), 520-27.

Bourne, R.B. *et al.* (1994) Ingrowth surfaces: plasma spray coating to titanium alloy hip replacements, *Clin Orthop Rel Res*, **298**, 37-46.

Brinker, M.R. *et al.* (1994) Primary total hip arthroplasty using noncemented porous-coated femoral components in patients with osteonecrosis of the femoral head, *J Arthroplasty*, **9** (5), 457-468.

Burke, D. (1985) *Advanced Concepts in Total Hip Replacement*, Ch. 3. (ed Harris) Slack Inc.

Charnley, J. (1972) The long term results of low friction arthroplasty of the hip performed as a primary intervention. *J Bone Joint Surg*, **54B**(1), 61-76.

Djerf, K. and Wahlstrom, O. (1986) Total hip replacement comparison between the McKee-Farrar and Charnley prostheses in a 5-year follow-up study. *Arch Orthop Trauma Surg*, **105**, 158-62.

Djerf, K. and Wahlstrom, O. (1986) Loosening 5 years after total hip replacement. *Arch Orthop Trauma Surg*, **105**, 339-42.

Dobbs, H.S. (1980) Survivorship of total hip replacements. *J Bone Joint Surg*, **62-B** (2), 168-73.

Furlong, R.J. and Osborn, J.F. (1991) Fixation of hip prostheses by hydroxyapatite ceramic coatings, *J Bone Joint Surg*, **73-B** (5), 741-45.

Jantsch, S. *et al.* (1991) Long-term results after implantation of McKee-Farrar total hip prosthesis, *Arch Orthop Trauma Surg*, **110,** 230-37.

Jayson, M. (1971) *Total Hip Replacement*, Sector Publishing, London.

Kelley, S. S. (1994) High hip center in revision arthroplasty, *J Arthroplasty*, **9**(5), 503-10.

McKee, G.K. and Farrar, J.W. (1966) Replacement of arthritic hips by the McKee-Farrar prosthesis, *J Bone Joint Surg*, **48B** (2), 245-59.

McKee, G.K. (1982) Total hip Replacement - past, present and future, *Biomaterials*, **3**, 130-35.

McKee, G.K. and Chen (1973) The statistics of the McKee-Farrar method *Clin Orthop Rel Res*, **95**, 26-33.

Miller, J. (1985) *Advanced Concepts in Total Hip Replacement,* Ch. 4, (ed Harris) Slack Inc.

Muller, M.E. (1992) Lessons of 30 years of total hip arthroplasty, *Clin Orthop Rel Res,* **274**, 12-21.

Oonishi, H. (1991) Orthopaedic applications of hydroxyapatite, *Biomaterials* **12**, 171-78.

Pepten, J.B. *et al.* (1989) Survivorship analysis of failure pattern after revision total hip arthroplasty, *J Arthroplasty,* **4** (4), 311-17.

Schutzer, S.F. and Harris, W.H. (1994) High placement of porous-coated acetabular components in complex THA, *J Arthroplasty,* **9** (4), 359-67.

Tillberg, B. (1982) THA using the McKee & Watson-Farrar prosthesis, *Acta Orthop Scand,* **53**, 103-7.

Visuri, T. *et al.* (1994) Life expectancy after hip arthroplasty, *Acta Orthop Scand,* **65** (1), 911.

Visuri, T. (1987) Long-Term results and survivorship of the McKee-Farrar total hip prosthesis, *Arch Orthop Trauma Surg,* **106**, 368-74.

Zaoussis, A.L. *et al.* (1989) Experience with total hip arthroplasty in Greece, the first 20 years, *Clin Orthop Rel Res,* **246**, 39-47.

6

A Comparison of Artificial Knee Arthroplasties

Penelope Kao
Shannon Eggers
Neil Graf
Bernd Liesenfeld

INTRODUCTION

The development of total knee arthroplasties (TKAs) closely followed that of total hip arthroplasties (THAs). However, knee replacement is a very different procedure from that of hip replacement. The knee is an inherently unstable joint and is subjected to greater applied loads since it is located more distally in the body. A comparison of the hip and knee implantation sites is illustrated in Table 6.1. Today, approximately 80,000 knee prostheses are implanted in the United States per year (Rand, 1993a). The success rates of TKAs currently equals or surpasses that of THAs. Surgeons predict a survivorship of 90% to 95% at 10 to 15 postoperative years for the TKA (Rand, 1993a).

The primary indication for performing a total knee arthroplasty is pain and severe restriction of routine daily activities (Rand, 1993a; Laskin, 1991). These patients should have exhausted all non-surgical forms of treatment and have radiographic evidence of severe joint destruction before considering knee arthroplasty (Rand, 1993a). The patient's vocation, age, weight and recreational activities are important criteria when selecting a TKA candidate. Since the success of the procedure depends on patient selection, implant choice and surgical technique, each contribution is of equal importance.

Table 6.1 Comparison of hip and knee implantation sites

Hip	*Knee*
Natural joint:	**Natural joint:**
Deep tissue location	Superficial tissue location
'Ball-in-socket' geometry	'End-on-end' geometry
Intrinsic stability	Stability depends on soft tissues
Single center of rotation	Combined rolling, sliding, and axial torsion
Free motion in all planes	Constrained, screw-home motion
Compressive stress	Compressive and impact stresses
Replacement prosthesis:	**Replacement prosthesis:**
Excess PMMA easy to remove during insertion	PMMA may be trapped posteriorly
Mimics natural structure	Menisci absent, capsule and ligaments may be damaged
Intrinsic stability	Stability depends in part on soft tissues
Compressive, bending stresses	Compressive, shear and torsional (axial) stresses
Peak load: 2-3 x body weight	Peak load: 3-4 x body weight
Peak stress: ⟨ 20 MPa	Peak stress: ⟩ 20 MPa
Wear debris clear from joint	Wear debris may be trapped

Knee replacement is often performed in patients with rheumatoid arthritis when other surgical procedures or medical therapy have failed (Laskin, 1991). The demanding nature of this disease results in patients placing less stress on their joints. As a result, arthroplasties can be performed on younger rheumatoid arthritis patients and be expected to exhibit less mechanical failure. Unfortunately, these patients may be more susceptible to delayed wound healing and sepsis (Laskin, 1991).

Osteoarthritis is another common disease that may necessitate a TKA. Osteoarthritis can be induced by various means, including genetic predisposition and specific trauma. However, the general consequence is a concentration of forces beyond the load bearing ability of the cartilage and subchondral bone (Buckwalter *et al.,* 1990). In the treatment of osteoarthritis, knee replacement versus knee realignment (osteotomy) must be considered. Arthritis may progress after a successful tibial osteotomy and knee replacement may become necessary. TKAs performed after osteotomy versus those performed on previously untreated knees have comparable results (Laskin, 1991).

BACKGROUND

The history of total knee arthroplasties is relatively short, dating back less than one hundred years. In 1909, one of the first TKAs involved replacing the entire knee with an allograft. At the time, allografts for knee replacements proved to be unsuccessful, due to the necrosis of bone, poor antibiotics and poor surgical procedures (Laskin, 1991). As total knee replacements evolved, metallic endoprostheses were used to replace only one side of the joint surface. In the 1950s, MacIntosh and McKeever performed hemiarthroplasties of the tibial plateau. The shortcoming of these devices was their lack of fixation, which led to the eventual migration of these devices. Also in the 1950s, Waldius experimented with hinged prostheses that were acrylic and later modified to be constructed of Vitallium® (Laskin, 1991). Several designs developed from the Waldius design, including the Geupar. These uncemented hinged prostheses had problems with loosening and settling.

The introduction of polymethylmethacrylate (PMMA) bone cement for fixation of hip implants paved the way for similar fixation of knee prostheses. Gunston, in the early 1970s, received credit for the first non-constrained knee arthroplasty with cemented prosthetic components. Gunston's pioneering work with the polycentric cemented prosthesis set the precedent for successful knee arthroplasties. The drawbacks of this four component modular design, were increased incidences of loosening of one or more of the components, and the complexity of the surgical techniques. The modular prosthesis for the treatment of bicompartmental arthritis was subsequently abandoned for these reasons.

Later in the 1970s, the difficulties of the four component design were eliminated by linking the two femoral components together and the two tibial components together, resulting in the dual component system. Subsequent variations of the metal femoral component/polyethylene tibial component system are widely used today.

DESIGN CONSIDERATIONS

Materials: femoral component

The material considerations of TKAs largely emulate those of THAs and have undergone little or no change since the 1970s. The metallic femoral component is generally cobalt-chromium alloy (Co-Cr). Co-Cr renders better articulation for the ultrahigh molecular weight polyethylene (UHMWPE) than do titanium alloys and stainless steel alloys. Ti-6Al-4V titanium alloy is believed to suffer accelerated wear as a result of catalytic carrier inclusions present in the UHMWPE and increased vulnerability to fretting corrosion due to its relative softness. Stainless steel cast alloys do not possess the necessary strength, while forged alloys are more expensive and cause more wear on the UHMWPE than Co-Cr. Ceramic components have not significantly been used to date (Walker, 1989).

Materials: tibial component

The tibial component is composed of an UHMWPE articulating surface, with or without a metallic tibial tray. Ultrahigh molecular weight polyethylene is produced by the Ziegler-Natta catalyst system. The Ziegler-Natta catalyst is able to yield unbranched and stereospecific polymers at room temperature and atmospheric pressure. The resulting polyethylene has molecular weights ranging from 1^{10} (Marmor, 1988) to 10^{10} (Walker, 1989). To date, UHMWPE has proven to be the best polymer for load-bearing applications in metal-polymer wear pairs (Black, 1988).

Surgical technique is paramount, as in all arthroplasties. The prosthesis must be placed with the mechanical axis of the limb falling onto the resurfaced compartment, because a tilting off-axis can overload the opposite compartment and ruin the intact articulating surface. Durable fixation, either uncemented or cemented, must be achieved for the resurfacing tibial component. Tibial components composed solely of UHMWPE, without the benefit of a metal backing, have inferior performance for the following reasons. Firstly, the unconstrained UHMWPE is susceptible to cold-flow creep, which deforms it and can lead to increasing wear and eventual failure. Subsequently, the UHMWPE-only tibial components, with a thickness of less than 9 mm, have been found to have unacceptably high rates of loosening.

Metal backing improves the prosthesis by constraining the UHMWPE, but adds more than 2 mm to the implant thickness, necessitating a greater bone resection. The metal backed tray allows the insertion of a replacement articular surface component, in the case of UHMWPE wear, without any bone loss (Marmor, 1988; Kozinn, Mora and Scott, 1989; Bernasek, Rand and Bryan, 1988).

Wear of the UHMWPE tibial component occurs by a number of mechanisms. General wear occurs when material is removed from the UHMWPE surface by the articulating counterface. Third body wear occurs due to the interposition of wear debris or other hard particles, leading to accelerated surface degradation. These hard particles may be UHMWPE, bone chips, PMMA cement or particles of spalled metallic, porous coating. Fatigue wear occurs due to the cyclic nature of the loading on the tibial component. Creep also occurs and contributes to wear. Creep can occur through cyclic loading at stresses below the yield strength of the UHMWPE in post-yield plastic deformation, or through stresses exceeding the yield strength of the UHMWPE. The creep deformation of the tibial component will impede articulation and limit the range of motion, compromising prosthesis performance. Creep deformation is limited by the use of a metal-backed tray to support the UHMWPE (Laskin, 1991).

Fixation methods

The availability of polymethylmethacrylate bone cement for fixation of knee prostheses opened the door for the successful treatment of arthritic conditions of the knee (Laskin, 1991). The prostheses from the 1950s and 1960s, which had no fixation mechanisms, failed due to loosening, settling, or displacement of the devices. The introduction and refinement of TKAs using PMMA bone cement has improved the durability of fixation with few failures reported solely due to loosening. The detrimental effects of using PMMA bone cement are attributed to the elevated temperatures which arise when the methylmethacrylate polymerizes *in situ*. These temperatures, around 45° C, can lead to bone necrosis.

As an alternative to PMMA cement, porous coating of femoral and tibial prostheses for mechanical fixation was introduced in the late 1970s and early 1980s. The porous surface, which allows for bone ingrowth, can be either metallic or bioactive ceramic. Prostheses implanted with porous coatings

coatings require longer times for fixation since the bone must grow into the implant in order to provide fixation. The advantages and disadvantages of the different fixation methods have yet to be determined.

Designs

The goals of total knee replacement are to eliminate pain and restore function. Figure 6.1 shows the salient features of a 'generic' total knee replacement arthroplasty. This design shows a conforming, closely anatomical, distal femoral component used with a resurfacing patellar insert and tibial component (Walker, 1989).

The knee is probably the most complex and demanding anatomic joint of our body (Rand, 1993). The human knee has cartilage which acts as both an articulating surface and a shock absorber. As a result, the knee joint resists compressive and impact stresses, which can be 3 to 4 times that of body weight. The human knee also has muscles and ligaments, which provide dynamic and passive stability, respectively. Sliding, rolling and axial torsion are all possible motions for the tri-articular knee joint. Artificial

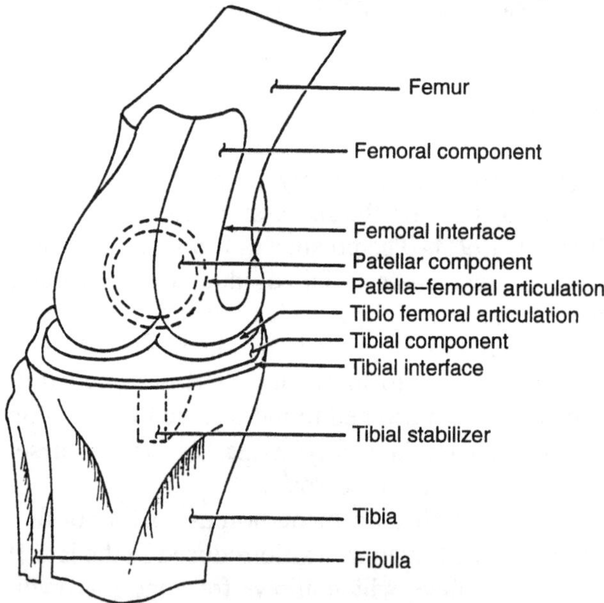

Fig. 6.1 Features of a generic total knee replacement arthroplasty.

surfaces can neither provide a natural range of motion, nor a large enough articulating surface needed to distribute stresses for a long lasting wear surface.

Most of the knee replacement prostheses presently used trace their lineage back to designs conceived in the early 1970s. For the purpose of this review, prostheses were grouped into 5 types according to basic joint design (Table 6.2). Some of these prostheses are only slight variations on a general design (Rand, 1993b; Marmor, 1988).

Table 6.2 Most commonly used knee prostheses

TYPE I HINGED
Waldius
Shiers
St. Georg
Stanmore
Guepar

TYPE III UNICOMPARTMENTAL	
McKeever	Oxford
MacIntosh	Lotus
Porous Coated Anatomic	Polycentric / Gunston Hult
Marmor	Unicondylar
Blauth	Robert Brigham
St. Georg Sledge	Mark I
Modular	Mark II
Geometric	

TYPE II CONSTRAINED
Total Condylar III
Kinematic Rotating Hinge
Sheehan
Attenborough
Spherocentric

TYPE IV MENISCAL BEARING
N. J. Low Contact Stress
Accord
Oxford Meniscal

TYPE V CONDYLAR RESURFACING

Poster Cruciate Ligament Retaining	Poster Cruciate Ligament Sacrificing	Poster Cruciate Ligament Substituting
Duocondylar Duopatellar Posterior Cruciate Condylar Kinematic I Kinemax Robert Brigham Townley / Anatomic Cloutier AGC Miller-Galante I Anatomic Modular Knee Press-Fit Condylar	Total Condylar Total Condylar II Freeman-Swanson ICLH Porous Coated Anatomic	Insall-Burstein Kinematic Stabilizer I Kinematic Stabilizer II Kinematic II

Type I: The hinged prosthesis

The hinged prosthesis was the first design to emerge and has since fallen almost completely into disuse (Fig. 6.2). The hinged prosthesis suffers from a very high failure rate due to loosening, because it transmits large torque loads and moments directly to the bone through the hinge. These large stresses exceed the interfacial strength of cements or other fixation methods. Early hinges suffered from abrupt impact-type loading when the hinge reached the end of its range of motion. More advanced designs have a more gentle stopper system. Hinges are presently only used in cases of extreme deformity and for extremely serious cases, where even the more advanced constrained designs do not offer sufficient stability.

Type II: The constrained (non-hinge) prosthesis

The constrained prosthesis is generally reserved for cases that require greater stability, such as severe varus or valgus alignment, or cases of more severe deterioration resulting in great laxity. Additionally, revision surgeries may also require the use of a constrained prosthesis. The additional constraint is provided by taller condyles on the femoral component, tracking on a more deeply grooved surface on the tibial side. However, this increased constraint

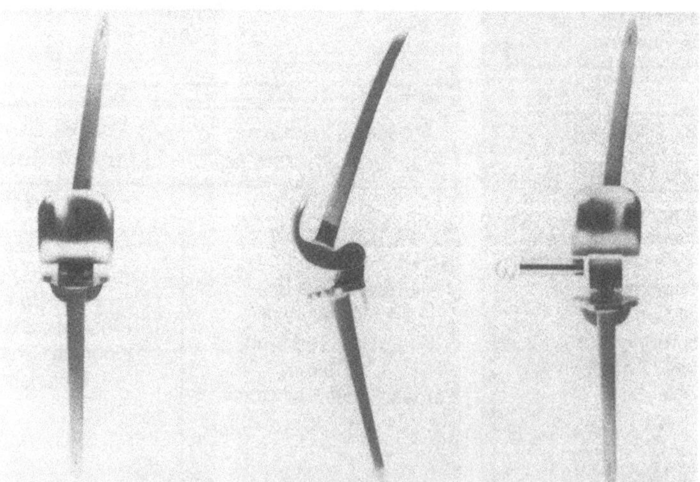

Fig. 6.2 Coronal, sagittal view and hinge mechanism of a typical hinge prosthesis. (From Rand 1993b, by permission of Mayo Foundation.)

compromises the range of motion that can be achieved and the normality of the gait. This increased constraint device also requires the resection of a larger volume of bone.

Type III: The unicompartmental prosthesis

The unicompartmental prosthesis resurfaces only a single tibial articular surface (Fig. 6.3). Typically, this refers to the medial compartment which is often the higher load bearing compartment. The cruciate ligaments, if still intact preoperatively, can be preserved to give the joint natural stability and range of motion. In addition, the implantation of unicompartmental prostheses is significantly less invasive. Since it requires the excision of less bone and tissue, i.e. the patellofemoral and opposite compartments, it is called 'bone preserving.' This minimally invasive surgery shortens the healing time, decreases the cost and significantly reduces the risk of infection which is the second leading cause for revision, loosening being the first.

Clinical data show that a unicompartmental arthroplasty is not suitable for patients with cartilage destruction due to inflammatory arthritis, such as rheumatoid arthritis, systemic lupus erythematosus, ankylosing spondylitis or psoriatic arthritis.

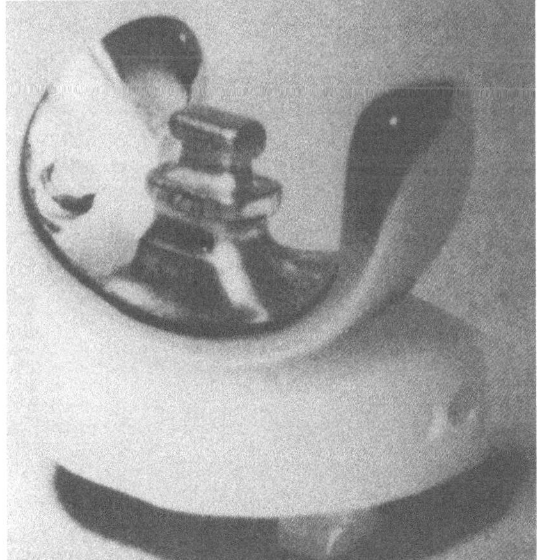

Fig. 6.3 The Unicompartmental (Unicondylar) Prosthesis. (From Rand 1993b, by permission of Mayo Foundation.)

Type IV: The meniscal bearing prosthesis

The meniscal bearing prosthesis attempts to more closely approximate the natural joint by emulating the mobile bearing mode of articulation, which is naturally provided by the synovial fluid (Fig. 6.4).

Type V: The condylar resurfacing prosthesis

Condylar resurfacing knee replacement prostheses constitute by far the largest group (Fig. 6.5). This group can be divided further into three subgroups according to the cruciate ligament treatment. Since the cruciate ligaments stabilize the knee joint, retention of the posterior cruciate ligament (PCL) will increase the stability of the joint without employing excessive mechanical constraint. As a result, PCL-retaining prostheses are believed to offer improved range of motion over designs that resect the ligament, such as the PCL-sacrificing and the PCL-substituting designs. The price of this improved motion is increased tibiofemoral contact stress, which may lead to faster wear and UHMWPE failure. The surgical procedure for retaining and properly aligning the ligament is more technically demanding than

Fig. 6.4 A meniscal bearing prosthesis. (From Rand 1993b, by permission of Mayo Foundation.)

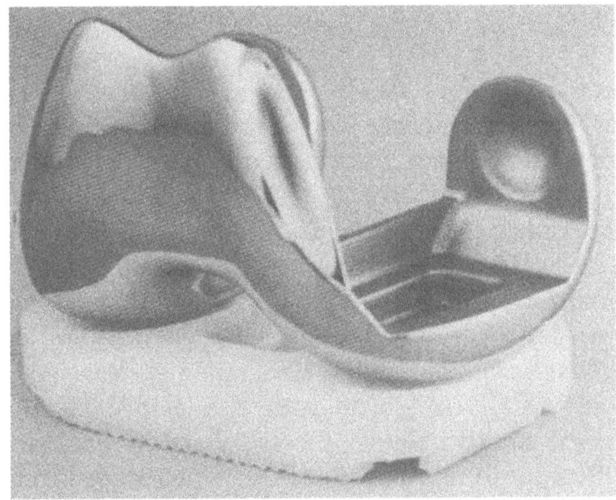

Fig. 6.5 The condylar prosthesis. (From Rand 1993b, by permission of Mayo Foundation.)

resecting the ligament, resulting in the relatively greater popularity of the PCL-sacrificing designs (Rand, 1993b).

The most popular implant available is the total condylar prosthesis, designed, at the Hospital for Special Surgery (HSS) in New York, by Insall, Ranawat and Walker. This PCL-sacrificing prosthesis was first implanted in 1974 and remains, with few modifications, the gold standard for knee arthroplasties. Originally designed to use PMMA cement fixation, the Total Condylar featured an all UHMWPE tibial platform. The tibial component was later modified in the late 1970s to include a metal backing. The porous-coated anatomical prosthesis (PCA) is designed for cementless fixation, but may be used with cement. Since resecting the ligament is popular, the PCA is generally used as a PCL-sacrificing prosthesis.

The PCL-substituting prosthesis, also known as posterior constrained prosthesis type, incorporates a small cam within the tibial articular surface to increase the rollback of the femoral component. This provides stability at higher angles of flexion.

The PCL-retaining prosthetic device, for example the Miller-Galante I prosthesis, preserves the posterior cruciate ligament. It features shallow, lightly anatomically inclined condyles (the medial being larger than the lateral), to improve the conformity with joint anatomy.

RESULTS

The purpose of this chapter was to compile data on all existing total knee arthroplasties. We were interested in general trends of the data. Frequently, the TKA reviews used either different scoring methods to rate the success of an implant or did not fully specify the implants used in the study. Since scoring is the absolute technique in determining successes and failures, the scoring must be based on the same criteria for the comparisons to be valid.

The Hospital for Special Surgery has a knee rating system which many investigators used to analyze their surgical data. Therefore, we opted to use only the data that was based on the HSS rating system. The HSS system uses a 100 point scale. Assessment categories include pain, function, range of motion, muscle strength, flexion deformity and instability. According to HSS, an excellent result requires 85 to 100 points, a good result requires 70 to 84 points, a fair result requires 60 to 69 points, and a poor result is below 60 points (Ranawat, Insall and Shine, 1976).

The results of the comparison indicate a great deal of similarity between the different types of implants, the data for the subsequent figures is presented in Table 6.4. Figure 6.6 displays the HSS 'good' and 'excellent' results (%) for all of structural types: hinged and constrained, unicompartmental, meniscal and condylar resurfacing knee arthroplasties. In this comparison, 1278 hinged and constrained arthroplasties, 3955 unicompartmental prostheses, 455 meniscal implants and 9191 condylar resurfacing knee implants were used to calculate the average of 'good' and 'excellent' percentages.

The average of the 'good' and 'excellent' percentages range from 82% to 90%, with meniscal implants having the highest percentage and unicompartmental having the lowest percentage of 'good' and 'excellent' scores. The hinged and constrained implants received 84%, while condylar resurfacing received 88%. The average follow-up times for the hinged and constrained implants ranged from 3 to 5 years; the unicompartmental implants ranged from 2 to 11 years; the meniscal arthroplasties ranged from 4.5 to 7.6 years; and the condylar resurfacing ranged from 1 to 15 years.

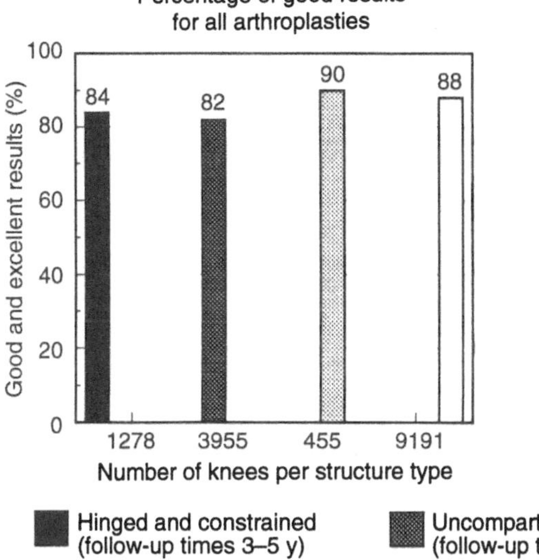

Fig. 6.6 Comparative results from total knee arthroplasties.

Figure 6.7 displays the percentage of 'good' and 'excellent' results as a function of follow-up time. The hinged and constrained plot has 4 data points (Fig. 6.7a) and the meniscal plot has 3 data points (Fig. 6.7c). Therefore, these plots carry little statistical significance due to the few data points. The plot of unicompartmental knee arthroplasties (Fig. 6.7b) indicates a decrease in 'good' and 'excellent' results as the postoperative time increases. The unicompartmental knees show an approximately 40% decline in 'good' and 'excellent' results 12 years after implantation. The plot of condylar knee prostheses (Fig. 6.7d) indicates a flat trend in the percentage of 'good' and 'excellent' results for up to 15 years after implantation. The scatter in the data points result from surgical technique, patient selection and subcategories of implants.

The subcategories for the condylar TKAs are shown in Fig. 6.8. Three subclassifications are presented: posterior cruciate ligament retaining, posterior cruciate ligament sacrificing and posterior cruciate ligament substituting. The percentages of 'good' and 'excellent' results are 88%, 84% and 96%, respectively. According to these plots, the type of condylar resurfacing implant does affect the results slightly.

Figure 6.9 is a comparison of the postoperative range of motion in degrees for the four main knee arthroplasties. The data indicate that hinged

Table 6.4 Data for Figs. 6.6-6.10

Reference		*Implant*	*Year of Surgery*	*Initial*	*Number of Knees*		*Used*	*Follow-up (yrs)*		*Age (yrs)*	
					Death	*N.U.*		*Avg*	*Range*	*Avg*	*Range*
Type I: Hinged											
Knutson	1986	Guepar, St. Georg, *et al.*	1975–83	410	0	0	410	3	1–6		
Rand	1991	Guepar, Walldius, *et al.*	1971–87	356	0	0	356	5	2–10	69	59–79
Type II: Constrained											
Knutson	1986	Attenborough, Spherocentric	1975–83	398	0	0	398	3	1–6		
Porter	1988	Sheehan	1978–82	85	0	25	60	3.5	2–6	63	41–81
Rand	1991	Total Condylar III, *et al.*	1971–87	114	0	0	114	5	2–10	70	58–82
Type III: Unicompartmental											
Kozinn	1989	Cemented (PE/metal)		55	0	0	55	5.5	4.5–6		
Bernasek	1988	PCA, Cementless	1984–85	33	4	1	28	2		64	51–80
Bensadoun	1989	Marmor & Cartier		100	0	8	92		2–12	69	
Blauth	1990	Blauth, Cemented (PE/metal)	1972–89	603	68	38	497	3.75	1–15	70	40–92
Capra	1989	Compartmental II		33	0	0	33	6.3	4–14	62.8	
Capra	1989	Marmor		19	0	0	19	11.1	4–14	62.8	
Insall	1980	Unicondylar, Cemented (PE)	1972–74	32	6	4	22	6	5–7	66	47–50
Knutson	1986	Paired Medial/Lateral	1975–83	1021	0	0	1021	3	1–6		
Knutson	1986	Medial	1975–83	2165	0	0	2165	3	1–6		
Knutson	1986	Lateral	1975–83	402	0	0	402	3	1–6		
Laurencin	1991	Brigham & Unicondylar	1979–87	24	0	1	23	6.75	3–12.75	67	
Marmor	1988	Marmor		60	0	0	60	10			
Lewallen	1984	Polycentric (PE)	1970–71	209	0	0	209	10		56.4	20–82
Marmor	1993	Marmor, Cemented (PE)	1975–90	34	0	0	34	5.5	2–16	68	60–84
Mink	1989	Marmor & Richards, Medial	1975–86	70	0	0	70				54–82
Mink	1989	Marmor & Richards, Lateral	1975–86	6	0	0	6				54–82
Page	1989	Zimmer & Marmor	1973–83	79	0	0	79	8		66	41–82
Rand	1991	Geom., Polycent (PCA)	1971–87	676	0	0	676	5	2–10	69	61–77
Rand	1991	Geometric & Polycentric (PE/metal)	1971–87	3159	0	0	3159	5	2–10	63	51–75
Scott	1981	Mark I & II Unicondylar	1974–80	100	0	0	100	3.5	2.6	71	44–85
Type IV: Meniscal Bearing											
Buechel	1989	N.J. LCS, Cemented (PE/metal)		149	0	0	149	7.6	2–10		
Buechel	1989	N.J. LCS, Cementless (PE/metal)		208	0	0	208	4.5	2–7.5	60	21–86
Johnson	1993	Accord, Cemented (PE/metal)	1982–85	133	29	6	98	5.75	4.5–8.2	69	31–87

Women	*Men*	*Diagnosis* *OA*	*RA*	*Other*	*Range of Motion* *Preop*	*Postop*	*HSS Score* *Preop*	*Postop*	*Results* *%Success*	*#Failures*
		140	270	0					80%	84
253	103	228	124	4					84%	77
		102	296	0					91%	43
		17	24	3	88	95				11
70	44	84	30	0					81%	19
				55					92%	0
12	13	25	0	3	112	119	66	81	74%	11
		73	0	27					83%	15
					97					24
		0	0						90%	
		0	0						54%	
25	5							48	31%	10
		276	745	0					88%	124
		2051	114	0					94%	140
		294	108	0					90%	39
10	13	22	1	0	106	123			96%	
									63%	21
107	52	54	140	15					62%	71
22	10	0	0	34	117	114	47.5	85.7	89%	4
					116	120			88%	3
					116	120			83%	1
					112	118			92%	16
352	324	656	13	7					86%	149
2053	1106	1579	1548	32					88%	739
67	33	94	2	4	112	113			92%	3
					92	109			85%	22
124	84	118	39	51	98	106			92%	17
106	27	59	74	0		100			93%	4

Table 6.4 Data for Figs. 6.6–6.10 (*Continued*)

Reference		*Implant*	*Year of Surgery*	*Initial*	*Number of Knees* *Death*	*N.U.*	*Used*	*Follow-up (yrs)* *Avg*	*Range*	*Age (yrs)* *Avg*	*Range*
Type V(A): Condylar Resurfacing – Posterior Cruciate Ligament Retaining											
Ranawat	1976	Duocondylar	1976	109	5	10	94	3	2–4	65	25–72
Bourne	1990	Miller–Galante, Cemented (PE)		60	0	0	60				
Bourne	1990	Miller–Galante, Cementless (PE)		60	0	0	60				
Cloutier	1983	Cloutier, Unconstrained (PE/metal)	1977–80	110	3	0	107	3	2–4.5	58	35–81
Dennis	1989	Posterior Cruciate Condylar		33	0	0	33	10		64	
Kirk	1994	Miller–Galante		50	4	1	45	2		69	56–80
Kirk	1994	Anatomic Modular Knee (PE)		50	2	0	48	2		67	41–76
Laurencin	1991	Duopatellar & Kinematic	1979–87	13	0	0	13	7.8	3–12.75	68	
Laurencin	1991	Kinemat. & Press-Fit Condyl.	1979–87	11	0	1	10	5.5	3–12.75	66.5	
Lee	1990	Posterior Cruciate Condylar (PE)	1975–81	315	109	59	144	8.9	7–13.3	67.7	22–88
Maloney	1992	Posterior Stabilized TC	1982–86	53	0	0	53	1.75		68	32–96
Rand	1993	Posterior Cruciate Condylar (PE/metal)		129	0	0	56	10	8–11.5	64	29–78
Rand	1993	Posterior Cruciate Condylar (PE)		61	0	0	22	10	8–11	58	27–71
Rand	1991	Kinematic, Townley, *et al.* (PE/metal)	1971–87	3907	0	0	3907	5	2–10	69	59–79
Rand	1991	Miller–Galante , Cementless	1971–87	310	0	0	310	5	2–10	56	45–67
Ritter	1989	Posterior Cruciate Condylar	1975–83	440	25	21	394	4.75	.75–10	69.8	32–91
Rosenberg	1989	Miller–Galante . Cemented (PE/metal)		133	0	0	133	2	0–4	70	31–97
Rosenberg	1989	Miller–Galante . Cementless (PE/metal)		134	0	0	134	2	0–4	58	19–75
Scott	1989	Press-fit Condylar (PE/metal)	1984–88	96	0	0	96	2		59	29–89
Townley	1985	Townley PCA, Cementless (PE/metal)	1972–85	532	0	0	532		1.5–11		
Volatile	1986	Duocondylar, Cemented	1973–75	171	45	63	64	10		62	
Windsor	1989	Post. Stabilized, Cemented (PE)	1974–86	289	0	0	289	10			
Windsor	1989	Post. Stablilized, Cemented (PE/metal)	1974–86	917	0	0	917	7			
Type V(B): Condylar Resurfacing – Posterior Cruciate Ligament Sacrificing											
Apel	1991	Total Condylar (PE)		62	0	0	62	7.5	2–11.4	61.5	26–81
Apel	1991	Total Condylar (PE/metal)		69	0	0	69	5.8	2–8	65.3	32–85
Insall	1979	Total condylar (PE)						6.6			
Insall	1979	Total Condylar (PE)							10–12		
Lee	1990	Total Condylar (PE)		144	0	0	144	8.9	7–13.3		
Collins	1991	PCA, Cemented	1983	25	0	0	25	3	2–5	65.6	46–81
Collins	1991	PCA, Cementless	1983	26	0	0	26	3	2–5	65.5	29–81
Cheng	1988	PCA, Cementless		43	0	0	43	1	0.5–3		
Hungerford	1985	PCA, Cementless		37	0	0	37				
Hungerford	1986	PCA, Cementless		93	0	0	93	3.25	2–5.4		
Hungerford	1987	PCA, Cementless		50	0	0	50		4–6.5		

		Diagnosis			Range of Motion		HSS Score		Results	
Women	*Men*	*OA*	*RA*	*Other*	*Preop*	*Postop*	*Preop*	*Postop*	*%Success*	*#Failures*
90	19	28	81	0	101	102			75%	8
									96%	3
									88%	7
63	20	35	68	4	95	104	38.3	86.1	91%	10
		16	17	0				84	91%	2
30	20	50	0	0		112		87		3
33	17	50	0	0		110		86		6
9	4	12	1	0	104	109			92%	1
6	4	10	0	0	113	113			70%	1
63	30	103	30	11	115	106	55	88	94.5%	8
23	14	24	9	4	102	108				
29	17	39	16	1	115	100	56	83	95%	2
9	8	11	11	0	115	102	65	84	86%	1
2344	1563	3047	820	40					98%	65
158	152	223	84	3					93%	8
282	158	334	92	14					86%	15
83	30	109	20	4	101	104	47	89	93%	7
74	52	108	19	7	103	107	51	91	95%	7
		67	29	0		113			90%	10
		388	95	49					89%	21
		41	129	2	86	103			44%	35
									97%	6
									99%	7
45	17	39	21	0		94.4		86.4		
49	20	49	20	0		96		88.3		
									91%	
									87.8%	
									94.5%	
18	7	12	10	3	103		57.6	81.5	68%	
22	4	16	9	1	106		59.6	80.4	69%	1
								83	78%	
									91%	
									94.5%	
								91	96%	

Table 6.4 Data for Figs. 6.6–6.10 (*Continued*)

Reference	*Implant*	*Year of Surgery*	*Number of Knees* *Initial*	*Death*	*N.U.*	*Used*	*Follow-up (yrs)* *Avg*	*Range*	*Age (yrs)* *Avg*	*Range*
Katz 1987	PCA, Cementless		15	0	0	15				
Rand 1986	PCA, Cementless		41	0	0	41				
Rorabeck 1988	PCA, Cementless		50	0	0	50		2–3		
Daluga 1987	PCA, Cemented		126	0	0	126	2.5	1.2–4.6		
Hungerford 1985	PCA, Cemented		70	0	0	70				
Katz 1987	PCA, Cemented		6	0	0	6				
Rand 1986	PCA, Cemented		50	0	0	50				
Ebert 1989	PCA, Cemented	1980–84	43	0	0	43		4–8	67	
Ebert 1989	PCA, Cementless	1980–84	78	0	0	78		4–8	55	
Maloney 1992	Total Condylar	1982–86	51	0	0	51	2.8		63	24–87
Mont 1994	PCA	1980–89	80	7	0	73	6.1	2–11	62	29–79
Ranawat 1989	Total Condylar, Cemented	1979–80	100	0	0	100	6.4	1–9	65.5	38–95
Rand 1991	Total condylar, *et al.* (PE)	1971–87	337	0	0	337	5	2–10	66	54–68
Rorabeck 1988	PCA, Cementless	1984–84	50	0	0	50	2.5	2–3		
Windsor 1989	Total Condylar, Cemented (PE)	1974–86	224	0	0	224	15			
Type V (C): Condylar Resurfacing – Posterior Cruciate Ligament Substituting										
Dennis 1992	PCL Substituting (PE)	1975–78	64	18	4	42	11	9.3–13	62.8	35–82
Rand 1991	Kinematic Stabilizer, *et al.*	1971–87	234	0	0	234	5	2–10	67	56–78
Rorabeck 1988	Kinematic II, Cemented	1984–85	110	0	0	110	2.5	2–3		

		Diagnosis			*Range of Motion*		*HSS Score*		*Results*	
Women	*Men*	*OA*	*RA*	*Other*	*Preop*	*Postop*	*Preop*	*Postop*	*%Success*	*#Failures*
								92	100%	
								83		
								79		
								87.5	90%	
									87%	
								95	100%	
									97%	
		0	43	0		94			80%	9
		0	78	0		103			93%	5
28	23	21	12	5	90	99				
36	37	65	8	0	95	102			64%	26
51	24	43	32	0		109.2	44.2		99%	1
216	121	219	115	3					94%	38
						110		78		8
									90%	12
		21	21	0	96	102.4		84.5	92.9%	7
152	82	133	98	3					96%	9
						110		88		3

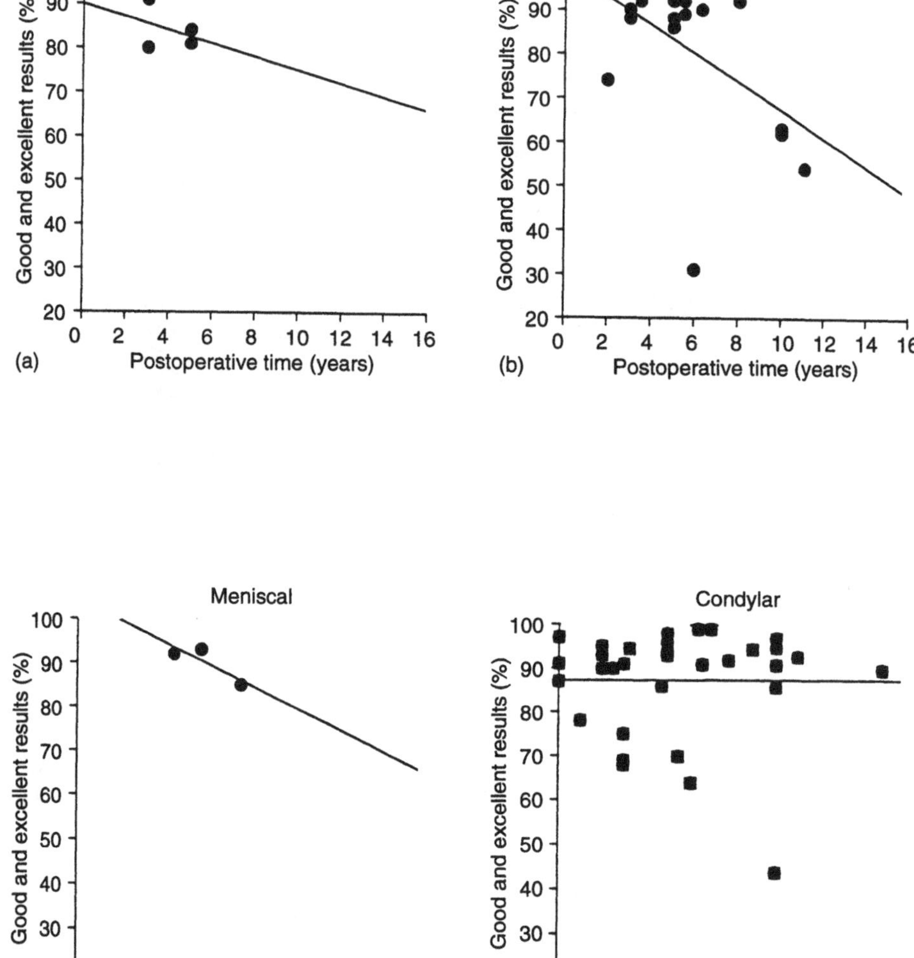

Fig. 6.7 Time dependence of clinical success of various types of TKA.

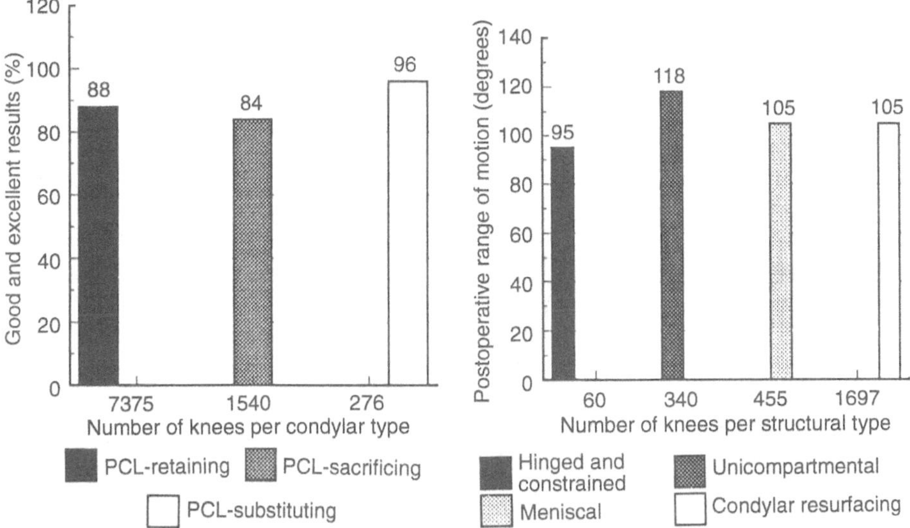

Fig. 6.8 Percentage of good results for condylar resurfacing arthroplasties.

Fig. 6.9 Range of motion for all arthroplasties.

and constrained arthroplasties have a 95° range of motion, the unicompartmental arthroplasties have a 118° range, the meniscal and condylar prostheses have a 105° range. These data imply that the hinged and constrained prostheses severely restrict the range of motion, since the average range of motion is approximately 125° in healthy individuals. The unicompartmental prosthesis provides the greatest amount of motion, probably due to the fact that it only replaces only one condylar surface. To be expected, the meniscal and condylar range of motions lie between the two previously mentioned prostheses.

Figure 6.10 depicts a comparison of the percentage of 'good' and 'excellent' results between PMMA cemented and cementless TKAs. The cemented TKAs scored 93% 'good' and 'excellent' results, while the cementless TKAs scored 91% 'good' and 'excellent' results: 2465 cemented knees were examined and 1614 cementless knees were examined. There was no statistical difference between cemented and cementless TKAs.

The failure modes included UHMWPE tibial component wear, extensor mechanism and patellar problems, aseptic loosening, deep infection and instability.

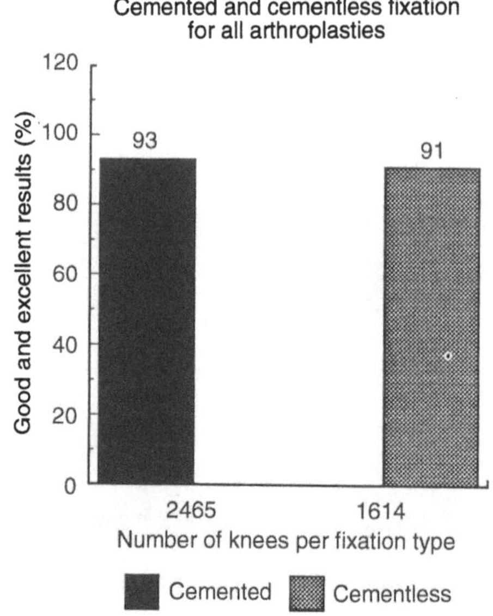

Fig. 6.10 Comparison of cemented vs cementless fixation of TKA.

CONCLUSIONS

Although most investigators used the HSS scoring system, which factored pain, function, range of motion, muscle strength, flexion deformity and instability, additional criteria are important for assessing the long-term success of a total knee arthroplasty. Success rates are higher for patients undergoing primary arthroplasty rather than revision arthroplasty, for patients with a diagnosis of rheumatoid arthritis rather than osteoarthritis, and for patients using a prosthesis with a metal-backed tibial component (Apel, Tozzi and Dorr, 1991). Unfortunately, these additional measurements were not mentioned or recorded by the majority of investigators.

The success rate of any implant is very hard to determine, let alone compare. Success depends not only on the implant. Surgical technique, patient selection and implant-patient match are all variables that warrant consideration. Consequently, the results presented in Figs. 6.6-6.10 can only be generally compared. In order for a comparison to be absolutely valid, each TKA must include patient history, rating system, surgical protocol and the specific knee prosthesis employed. It is unlikely that any comparison will ever be absolute.

REFERENCES

Apel, D.M., Tozzi, J.M. and Dorr, L.D. (1991) Clinical comparison of all-polyethylene and metal-backed tibial components in total knee arthroplasty. *Clinical Orthopaedics* **273**, 243.

Bensadoun, J.L., Vidal, J. and Salvan, J. (1989) Unicompartmental knee arthroplasty. *Orthopaedic Transactions* **13**, 708.

Bernasek T.L., Rand, J.A. and Bryan, R.S. (1988) Unicompartmental porous coated anatomic total knee arthroplasty. *Clinical Orthopaedics* **236**, 52.

Black, J. (1988) *Orthopaedic Biomaterials in Research and Practice*, Churchill Livingstone, New York.

Blauth, W. and Hassenpflug, J. (1990) Are unconstrained components essential in total knee arthroplasty. *Clinical Orthopaedics* **258**, 86.

Bourne, R.B., Rorabeck, C.H. and Nott, L. (1990) A prospective two-year comparison of the hybrid cemented and cementless Miller-Galante total knee replacement in osteoarthritic patients. *Proceedings and Reports of Universities, Colleges, Councils, Associations and Societies* **72-B**(3), 541.

Buckwalter, J.H. *et al.* (1990) *Articular Cartilage and Knee Joint Function* (ed J.W. Ewing) Raven Press, New York.

Buechel, F.F., and Pappas, M.J. (1989) New Jersey low contact stress knee replacement system. *Orthopaedic Clinics of North America* **20** (2), 147.

Cage, D.J., Neill, W., Granberry, M. and Tullos, H.S. (1992) Long-term results of total arthroplasty in adolescents with debilitating polyarthropathy. *Clinical Orthopaedics* **283**, 156.

Capra, S. and Fehring, T. (1989) Unicompartmental arthroplasty, a four to fourteen year review. *Orthopaedic Transactions* **13**, 550.

Cloutier, J.M. (1983) Results of total knee arthroplasty with a non-constrained prosthesis. *Journal of Bone and Joint Surgery* **65-A** (7), 906.

Collins, D.N. *et al.* (1991) Porous-coated anatomic total knee arthroplasty: a prospective analysis comparing cemented and cementless fixation. *Clinical Orthopaedics* **267**, 128.

Dennis, D.A. *et al.* (1989) Posterior cruciate condylar total knee arthroplasty. *Orthopaedic Transactions* **13**, 73.

Dennis, D.A. (1992) Posterior cruciate condylar total knee arthroplasty. Average 11-year follow-up evaluation. *Clinical Orthopaedics* **281**, 168.

Ebert, F. *et al.* (1989) A four to eight year follow-up of cemented versus uncemented arthroplasty in rheumatoid patients. *OrthopaedicTransactions* **13**, 550.

Hofmann, A.A., Beck, S.W. and Wyatt, R.B. (1989) Cementless total knee arthroplasty in patients over 65 years old. *Orthopaedic Transactions* **13**, 74.

Insall, J. and Agliett, P. (1980) A five to seven-year follow-up of unicondylar arthroplasty. *Journal of Bone and Joint Surgery* **62-A** (8), 1329.

Johnson, R. *et al.* (1993) Five to eight year results of the Johnson-Elloy (Accord) total knee arthroplasty. *Journal of Arthroplasty* **8** (1), 27.

Kirk, P.G. (1994) Clinical comparison of the Miller Galante I and AMK total knee systems. *Journal of Arthroplasty* **9** (2), 131.

Knutson, K., Lindstrand, A. and Lidgren, L. (1986) Survival of knee arthroplasties: a nation-wide investigation of 8000 cases. *Journal of Bone and Joint Surgery* **68-B** (5), 795.

Kozinn, S.C., Mora, C. and Scott, R.D. (1989) Unicompartmental knee arthroplasty: a 4.5-6 year follow-up study with a metal-backed tibial component. *Journal of Arthroplasty Supplement* **1**.

Laskin, R.S., Denham, R.A. and Apley, A.G. (1984) *Replacement of the Knee,* Springer-Verlag, New York.

Laskin, R.S. (1991) *Total Knee Replacement,* Springer-Verlag, London.

Laurencin, C.T. *et al.* (1991) Unicompartmental versus total knee arthroplasty in the same patient. A comparative study. *Clinical Orthopaedics* **273**, 151.

Lee, J.G. *et al.* (1990) Review of the all-polyethylene tibial component in total knee arthroplasty. A minimum seven-year follow-up period. *Clinical Orthopaedics* **260**, 87.

Lewallen, D.G., Bryan, R.S. and Peterson, L.F. (1984) Polycentric total knee arthroplasty. *Journal of Bone and Joint Surgery* **66-A** (8), 1211.

Maloney, W.J., and Schurman, D.J. (1992) The effects of implant design on range of motion after total arthroplasty. *Clinical Orthopaedics* **278**.

Marmor, L. (1993) Unicompartmental arthroplasty for osteonecrosis of the knee joint. *Clinical Orthopaedics* **294**, 247.

Marmor, L. (1988) Unicompartmental arthroplasty of knee with minimum ten-year follow-up period. *Clinical Orthopaedics* **228** 171.

Mink, W.F. (1989) Unicompartmental knee arthroplasty. *Orthopaedic Transactions* **13**, 73.

Mont, M.A. *et al.* (1994) Total knee arthroplasty after failed high tibial osteotomy. A comparison with a matched group. *Clinical Orthopaedics* **299**, 125.

Page, L.V. and Mullen, M.P. (1989) Results in unicompartmental knee arthroplasty. *Orthopaedic Transactions* **13**, 73.

Porter, M. and Hirst, P. (1988) The Sheehan total knee arthroplasty. A retrospective review. *Clinical Orthopaedics* **236**, 227.

Ranawat, C.S., Esall, J. and Shine, J. (1976) Duo-condylar knee arthroplasty: Hospital for Special Surgery design. *Clinical Orthopaedics* **120**, 76.

Ranawat, C.S. and Hansraj, K.K. (1989) Effect of posterior cruciate sacrifice on durability of the cement-bone interface. *Orthopaedic Clinics of North America* **20** (1), 63.

Rand, J.A. (1993a) Comparison of metal-backed and all-polyethylene tibial components in cruciate condylar total knee arthroplasty. *Journal of Arthroplasty* **8** (3), 307.

Rand, J. (1993b) *Total Knee Arthroplasty,* Raven Press, New York.

Rand, J.A. and Ilstrup, D.M. (1991) Survivorship analysis of total knee arthroplasty. *Journal of Bone and Joint Surgery* **73-A** (3), 397.

Ritter, M.A. *et al.* (1989) Long-term survival analysis of the posterior cruciate condylar total knee arthroplasty. *Journal of Arthroplasty* **4** (4), 293.

Rorabeck, C.H., Bourne, R.B. and Nott, L. (1988) Cemented Kinematic II and non-cemented PCA total knee joint replacement: Prospective comparison. *Orthopaedic Transactions* **21**, 654.

Rosenberg, A.G., Barden, R. and Galante, J. (1989) A comparison of cemented and cementless fixation with the Miller-Galante total knee arthroplasty. *Orthopaedic Clinics of North America* **20** (1), 97.

Scott, R.D. and Santore, R.F. (1981) Unicondylar unicompartmental replacement for osteoarthritis of the knee. *Journal of Bone and Joint Surgery* **63-A** (4), 536.

Scott, R.D. and Thornhill, T.S. (1989) Press-fit condylar total knee replacement. *Orthopaedic Clinics of North America* **20** (1), 89.

Scuderi, G.R., Insall, J. and Windsor, R.E. (1989) Survivorship of cemented knee replacements. *Journal of Bone and Joint Surgery* **71-B** 798.

Stuart, M.J., Larson, J.E. and Morrey, B.F. (1993) Reoperation after condylar revision total knee arthroplasty. *Clinical Orthopaedics* **286**, 168.

Stulberg, S. *et al.* (1988) Four year follow-up of a porous-coated unicompartmental total knee replacement: Correlation with an animal model. *Orthopaedic Transactions* **21**, 654.

Thornhill, T.S. and Scott, R.D. (1989) Unicompartmental total knee arthroplasty. *Orthopaedic Clinics of North America* **20** (2), 245.

Townley, C.O. (1985) The anatomic total knee resurfacing arthroplasty. *Clinical Orthopaedics* **192**, 82.

Volatile, T.B. and Ewald, F.C. (1986) Ten year results of 171 duocondylar total knee replacements. *Orthopaedic Transactions* **10**, 491.

Walker, P. (1989) Requirements for successful total knee replacements, design considerations. *Orthopaedic Clinics of North America* **20** (1), 15.

Walker, P. (1989) Requirements for successful total knee replacements, material considerations. *Orthopaedic Clinics of North America* **20** (1), 1.

Windsor, R.E. *et al.* (1989) Mechanisms of failure of the femoral and tibial components in total knee arthroplasty. *Orthopaedic Transactions* **13**, 550.

7

Shoulder Implant System

Greg Snyder

INTRODUCTION

The history of shoulder implants is generally believed to have its origins in late 19th century France. Designed by a Parisian dentist, J. Michael Porter, (Lugli, 1978) the Pean artificial shoulder was successfully used to reconstruct the shoulder of a patient who had been afflicted with a tubercular infection, causing purulent sores to form on the proximal humerus. This prosthesis would be classified as a constrained-type implant, where the humeral element is physically constrained by the glenoid portion of the implant, which is in turn attached to the scapula. Amazingly, the implant functioned for over two years, up to 1893, allowing the use of the affected arm. As will be seen from this chapter, this is a remarkable accomplishment, even today, for arguably the most complex joint in the human body.

The nature of shoulder arthroplasty is complicated by the number and variation of indications for which an implant could be considered. It is generally agreed that it is best to avoid surgery if possible. If surgery is required and an implant is deemed necessary, the decision regarding the type of implant to use can be a very difficult one.

The types of implants available for shoulder arthroplasty can be classified into one of the following categories:

1) unconstrained total arthroplasty;
2) semi-constrained total arthroplasty;
3) constrained total arthroplasty;
4) unconstrained hemi-arthroplasty;
5) bipolar shoulder replacement.

In general, the success rate of a given implant system is closely related to the overall condition of the shoulder into which it is placed. Furthermore, it appears that the fewer modifications that are required to the existing structure, the less the potential for complications.

In the following discussion, each implant system is described, and its clinical success is evaluated. In addition, the various failure modes and the impact of the preoperative condition of the joint upon prognosis will be discussed. The varying indications, lengths of follow-up periods and methodologies used by the clinicians to quantify success interfere with efforts to compare, objectively, one system with another.

IMPLANT SUCCESS/FAILURE

The failures and complications associated with shoulder replacement surgery are too numerous to mention, let alone describe in detail, in this chapter. The prevalent modes of failure of the above-described systems, and the potential complications associated with the surgery, healing and everyday use of these systems are addressed.

The condition of the shoulder that justifies use of an implant is, by its very nature, less than optimal. Indications for implant surgery include any one of many arthritic conditions, osteonecrosis, tumors, chronic dislocation, various types of fractures, as well as muscle, tendon and synovial tissue damage. The implant system chosen will be influenced by the condition of the humerus, glenoid and the connective tissues of the joint.

The classic mode of failure in just about any implant system is implant loosening. This can occur under a large number of different circumstances. One of the primary causes for implant loosening is misalignment of the load path of the joint. This can be caused by improper selection of humeral length, restrictions in available bone for the glenoid component, improper bone surface preparation, poor cement quality and abuse of the joint by the patient. Also the patient may fall or otherwise impact the joint, which may cause the implant/bone interface to loosen.

The most frequently mentioned cause for suboptimal shoulder implant performance is compromised rotator cuff tissue. This condition results in reduced stability of the joint as well as lost range of motion. The prognosis for restoration of the joint to normal function is not good when severe rotator cuff limitations exist. Often a constrained or semi-constrained prosthesis will be prescribed to improve stability of the repaired joint.

Dislocations and subluxations of the shoulder are common after shoulder replacement surgery. Creation of a stable shoulder joint is almost always done at the expense of range of motion. The unconstrained type of implant attempts to balance these desired characteristics. Unfortunately, the balance between the two is delicate, and it is not uncommon for subluxations and dislocations to occur. Under normal operation, the frequency of these occurrences decreases as the joint heals following surgery. In some patients however, the bone and muscle structure never regains its original stability and revision surgery is required. The glenoid component can be replaced with a larger component, or one with a 'hood' on the perimeter. In other cases a constrained implant is prescribed. In either case, range of motion is compromised.

Another somewhat common, very painful, affliction to the patient after shoulder arthroplasty is impingement of the remaining proximal humeral bone on the acromion. This condition is usually the result of the humeral implant component being placed too deep within the humerus, thus allowing the remaining bone to project above the shoulder of the implant. Surgery is often required to remove the surplus humeral bone.

As with all surgery, the risk of infection is always present. With proper procedure the risk of subsequent infection is minimized. However there are many cases reported where removal of shoulder implants was caused by infection.

The primary interoperative complication that regularly occurs in shoulder arthroplasty is perforation of the cortical bone of the humerus during preparation of the medullary canal for receipt of the stem component. With proper care to contain cement within this hole, this complication is unlikely to develop into a serious problem.

Based on the literature reviewed for this chapter, the most common causes of complications following implant surgery are brought about by external sources, not always controllable by the implant designer or the surgeon. Among these causes of failure are: accidental falls and trauma, improper therapy, participation by the patient in inappropriate activity, substance abuse and mental illness. Several cases have been reported where a recipient, with low motivation and low threshold to pain, complained of symptoms that could not be attributed to any known implant related causes.

Revision surgery was not attempted, and the disposition of these cases as successes or failures was uncertain.

Definition of implant failure

The goal of this chapter is to evaluate the relative merits and liabilities of the various implant systems available for the shoulder. With this goal in mind, failure of an implant will be acknowledged when the implant, or its attachment to the body, is compromised in such a way that replacement or removal is required. If the removal is because of surgical error, the possibility of this complication will be noted but it will not be categorized as a failure of the implant. Accidents and abuses which result in failure of the implant will be treated similarly.

Definition of implant success

Success of an implant is relative. For a patient with no function, or extreme pain, the presence of function or absence of pain is a tremendous benefit. The issue of success becomes clouded when relative degrees of pain, mobility and functionality must be assessed.

The following methods have been used by clinicians to evaluate the success of various implant systems. Early in clinical trials, evaluation of success tended to be more subjective. These definitions of success are presented here in ascending order of objectivity.

Relative Pain
None, Mild, Moderate, Severe

Measured Range of Motion
Active Abduction
External Rotation
Internal Rotation
Flexion
Extension

Relative Range of Motion

No movement, < 1/3 range, < 2/3 range, > 2/3 range

Ability to Perform Common Daily Activities

Dress
Eat using knife, fork, or spoon
Personal hygiene
Comb hair
Sleep on operated limb without significant pain
Lift moderate weight with limb near side
Use hand for light work with arm positioned at shoulder level

Radiographic Evaluation

Measurement of thickness of radiolucent line at interface

Subjective Patient Response

1) Unchanged, Better, Much Better
2) Excellent, Satisfactory, Unsatisfactory, Limited Goals

Excellent:	no major pain, full use of the shoulder for daily activities, nearly normal strength, forward elevation to within 35 degrees of normal, and 90% of the range of internal and external rotation of a normal shoulder.
Satisfactory:	occasional pain or pain with weather changes, full shoulder use at least below top of the head for daily activities, >30% normal muscle strength, 90-135 degrees forward elevation, >50% of the range of rotation of the normal side.
Unsatisfactory:	did not meet above criteria.
Limited Goals:	used for patients with poor prognosis and usually with a history of failed procedures.

Some clinicians have attempted to standardize the evaluation of implant success. At this time there is no consensus upon which method to use. Several examples of standard evaluation forms are shown in Tables 7.1-7.3.

It can be seen from these criteria that the definition of success is far more arbitrary than that of failure. The direct comparison of results, using these various criteria, is in itself an arbitrary exercise. In addition to the differences in evaluation criteria, each case is different from all others in the following respects:

Age of the patient
Health of the patient
Lifestyle of the patient
Sex of the patient
Indications

It can be argued that the best measure of success is the relative improvement of patient lifestyle, postoperative comparison to normal ability of the joint or maybe most important, and most subjective, satisfaction of patient in their overall condition.

In addition to discussion of success and failures, the rate of occurrence of complications can be an indication of the relative robustness of an implant design. Regardless of the number of good results, any number of questionable or poor results is a source of distress to the clinician, and more importantly, the affected patient. The most significant complications are those which occur in a supposedly standard application. These will be noted in the discussion of the clinical histories of the implant systems.

ANALYSIS

Even though shoulder implants have existed since the late 19th century, only recently have materials and procedures developed to a degree where widespread clinical use can occur. Due to the relative lateness of these developments, the longest evaluation of any system found in the literature was approximately 10 years. Many of the systems examined had clinical trial results for follow-up periods of only 2 or 3 years. The differences in the follow-up periods of the various trials will be taken into account in their evaluation.

Table 7.1 American shoulder and elbow surgeons shoulder evaluation form

Name__________ Hosp. #__________ Date__________ Shoulder R/L

I. PAIN: (5 = none, 4 = slight, 3 = after unusual activity, 2 = moderate, 1 = marked, 0 = complete disability, NA = not available) _____

II. MOTION:

A. Patient Sitting

1. Active total elevation of arm: __________degrees*
2. Passive internal rotation:
   (Circle segment of posterior anatomy reached by thumb)
   (Note if reach restricted by limited elbow flexion)

1 = Less than trochanter	5 = L5	9 = L1	13 = T9	17 = T5
2 = Trochanter	6 = L4	10 = T12	14 = T8	18 = T4
3 = Gluteal	7 = L3	11 = T11	15 = T7	19 = T3
4 = Sacrum	8 = L2	12 = T10	16 = T6	20 = T2

3. Active external rotation with arm at side: __________degrees
4. Active external rotation at 90° abduction: __________degrees
   (Enter "NA" if cannot achieve 90° abduction)

B. Patient supine

1. Passive total elevation of arm: __________degrees*
2. Passive external rotation with arm at side __________degrees

*Total elevation of arm measured by viewing patient from side and using goniometer to determine angle between *arm* and *thorax*.

III. STRENGTH: (5 = normal, 4 = good, 3 = fair, 2 = poor, 1 = trace, 0 = paralysis)

A. Anterior deltoid __________ C. External rotation __________
B. Middle deltoid __________ D. Internal rotation __________

IV. STABILITY: (5 = NORMAL, 4 = APPREHENSION, 3 = rare subluxation, 2 = recurrent subluxation, 1 = recurrent dislocation, 0 = fixed dislocation, NA = not available)

A. Anterior __________ B. Posterior __________ C. Interior __________

V. FUNCTION: (4 = normal, 3 = mild compromise, 2 = difficulty, 1= with aid, 0 = unable, NA = not available)

A. Use back pocket	______	I. Sleep on affected side	______
B. Perineal care	______	J. Pulling	______
C. Wash opposite axilla	______	K. Use hand overhead	______
D. Eat with utensil	______	L. Throwing	______
E. Comb hair	______	M. Lifting	______
F. Use hand with arm at shoulder level	______	N. Do usual work	______
G. Carry 10-15 lbs. with arm at side	______	O. do usual sport	______
H. Dress	______		

Table 7.2 UCLA shoulder rating system

Findings	*Score (Points)*
Pain	
Constant and unbearable; patient needs strong medication frequently	1
Constant but bearable; patient needs strong medication occasionally	2
None or little at rest; occurs with light activities; patient needs salicylates frequently	4
With heavy or particular activities only; patient needs salicylates occasionally	5
Occasional and slight	8
No pain	10
Function	
Unable to use arm	1
Very light activities only	2
Light housework or most activities of daily living	4
Most housework, washing hair, putting on brassiere, shopping, driving	5
Slight restriction only; able to work with arm above shoulder level	8
Normal activities	10
Muscle power and motion	
Anklyosis with deformity	1
Anklyosis with good functional position	2
Muscle power, poor to fair, elevation <60°, internal rotation <45°	4
Muscle power, fair to good; elevation <90°, internal rotation <99°	5
Muscle power, good or normal; elevation, 140°, external rotation, 20°	8
Normal muscle power, motion nearly normal	10

Figure 7.1 shows the relative numbers of implantations for each of the reported indications. For these studies, rheumatoid arthritis is by far the most common indication in the prescription of a shoulder prosthesis. Arthritis, in general, is responsible for the implantation of nearly three quarters of the prostheses reported upon in the literature, as indicated in Table 7.4. The repair of various failed surgeries accounts for 4% of the reported implantations, with pre-existing old fractures and osteonecrosis of the humeral head being responsible for another 3.4% and 3.0%, respectively.

Neer reports (1985) that in 500 shoulder arthroplasties he performed up to 1985, 42% of the implants were prescribed for relief of arthritis, 36% of the implants were for trauma repair, both acute and chronic, 9% were

Table 7.3 Shoulder score system

ROM Score (10 points)

	Points	ROM
Abduction (2 points)	0.4	<20°
	0.8	21°-40°
	1.2	41°-60°
	1.6	61°-80°
	2.0	>80°
Adduction (1 point)	0.2	<10°
	0.4	11°-20°
	0.6	21°-30°
	0.8	31°-40°
	1.0	>40°
Extension (1 point)	0.2	0°
	0.4	1°-10°
	0.6	11°-20°
	0.8	21°-30°
	1.0	>30°
Flexion (4 points)	0.8	<20°
	1.6	21°-40°
	2.4	41°-60°
	3.2	61°-80°
	4.0	>80°
Internal rotation (1 point)	0.2	<20°
	0.4	21°-40°
	0.6	41°-60°
	0.8	61°-80°
	1.0	>80°
External rotation (1 point)	0.2	0°
	0.4	1°-10°
	0.6	11°-20°
	0.8	21°-30°
	1.0	>30°

Pain Score (10 points)

Degree	Points
Pain free	<10
Minimal pain after heavy work	8
Pain with daily activity	6
Pain with shoulder motion	4
Pain at rest	2

ADL Score (10 points)

Activity	Points
Independent, normal activities	10
Slight restrictions for heavy work overhead	8
Most ADL	6
Light activities only, assistance for some ADL	4
Inability to use shoulder for function	2

Shoulder Score (30 points)	Points
Poor	<18
Fair	18-22.9
Good	23-27.9
Excellent	28-30.0

ADL = activities of daily living

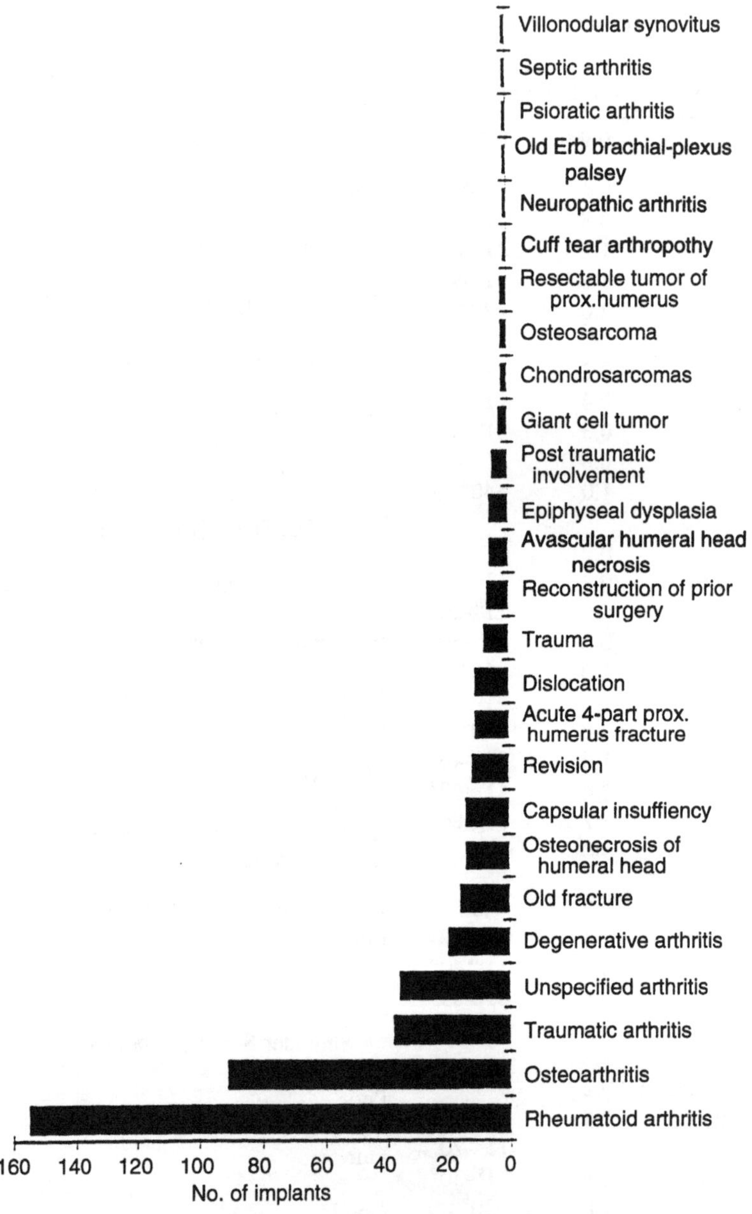

Fig. 7.1 Number of implants per clinical indication.

Table 7.4 Shoulder implant frequency, grouped by similar indications

Indication	*Cases*	*Studies*	*%Total*
Arthritis	**343**	**29**	**73.8**
Rheumatoid Arthritis	155	11	33.3
Osteoarthritis	91	7	19.6
Traumatic Arthritis	38	5	8.2
Unspecified Arthritis	36	1	7.7
Degenerative Arthritis	20	2	4.3
Septic Arthritis	1	1	0.2
Neuropathic Arthritis	1	1	0.2
Psioratic Arthritis	1	1	0.2
Repair of failed surgery	**19**	**4**	**4.0**
Revision	12	3	2.6
Reconstruction of prior surgery	7	1	1.5
Tumor	**9**	**4**	**1.9**
Giant Cell Tumor	3	1	0.6
Chondrosarcomas	2	1	0.4
Osteosarcoma	2	1	0.4
Resectable Tumor of Prox. Humerus	2	1	0.4

revisions of other surgeons' works, 5% for cuff-tear arthropathy and 4% for avascular necrosis. Since the different implant systems are often designed for correction of a specific condition, these percentages will vary for each system.

Unconstrained total shoulder arthroplasty

(a) System description

Possibly the most popular choice for shoulder replacement today, is the unconstrained total shoulder replacement. This class of implant generally consists of a metal humeral stem/ball component that rides on a ultrahigh molecular weight polyethylene (UHMWPE) surface that has been attached to the scapula. The glenoid component may be made of UHMWPE alone, ar it may have metal backing with a polyethylene sleeve. The bearing surface of this component is usually spherical, with a rather large radius. The most popular choices for materials for the humeral component are stainless steel and cobalt chrome alloys. The 'ball' end of this component is typically spherical or ellipsoid, with a radius of curvature less than that of the glenoid surface. A typical unconstrained implant is illustrated in Neer, 1985.

This type of implant allows the greatest potential for restoration of near-normal range of motion of the shoulder joint. It also appears to be somewhat tolerant of less-than-optimal installation, whether due to surgical error or poor condition of the implantation site. Selection of too long or too short a humeral component will result in superior or inferior subluxation, limiting the range of motion, or worse yet, requiring revision surgery due to pain, unreducable dislocation or prosthesis loosening. In cases of poor muscle and tendon development, there is a good chance that the humeral component will not be adequately constrained and a dislocation or subluxation may occur. Even if reduction is not necessary, or closed reduction is possible, loading upon the glenoid component may be less than optimal, resulting in premature failure of the glenoid component or its attachment to bone.

(b) Clinical results

Neer, (1988) 19 implants, 53 month average follow-up:
Of the papers cited, Neer presents the clinical evaluation of the longest follow-up period in which specific prosthesis performance evaluation is presented. Nineteen Neer implants were evaluated in patients with severe glenoid abnormalities which required bone grafts before implantation of the prosthesis. The severity of the condition of the glenoids of two of the patients resulted in a limited goals prognosis. Fourteen of the fifteen remaining patients achieved an excellent result over periods of 24-89 months. The remaining patient continued to experience occasional pain. In four patients, glenoid fixation screws broke. The patients remained asymptomatic, and the broken screws were discovered only after follow-up radiographs. Only excellent patient ratings were received in cases where metal-backed glenoid components were used. Neer suggests their use, unless specific reasons exist that are contraindications.

Failures 0, Complications 21%, no unsatisfactory results.

Amstutz *et al.,* (1988) 56 implants, 42 month average follow up:
This series of implants, the Dana implant, was designed at the University of California, Los Angeles (UCLA). The rate of failure after an average follow-up period of 42 months, was 4%, with 13% of the implants experiencing complications of some sort. The rate of complications for this system per total months of follow-up was the lowest of any group examined. The patient population was mixed, with arthritis and failed arthroplasty being the primary indications. The patients who experienced complications were those with tears of the rotator cuff. The post-traumatic arthritic group studied frequently required the implantation of 'hooded' glenoid components in order to provide additional stability. Impingement of the humeral component upon the glenoid occurred in this group when abduction exceeded 90 degrees, and caused the glenoid to loosen in one of ten such patients.

Failures 4%, Complications 13% (rotator cuff tear related).

Barrett, *et al.*, (1987) 50 implants, 42 month average follow up:
In this study at the University of Washington Medical School, Neer II implants were used in patients with arthritis and previous humeral head fractures. Nine of the shoulders examined had tears of the rotator cuff. After an average follow-up period of 42 months the failure rate was 6% and the rate of complications was 48%. This study had the highest rate of complications of any of the unconstrained total arthroplasty systems reported. Loosening of the glenoid occurred in shoulders with rotator cuff tears in 4 of the 5 cases that experienced loosening. Eight perioperative complications occurred.

Failures 6%, Complications 48% (interoperative, implant selection and rotator cuff related).

McElwain and English, (1987) 13 implants, 37 month average follow-up:
This Canadian study examined porous coated humeral and glenoid components of the author's design. The failure and complication rates were not noticeably different from the rest of the studies, although the length of the follow-up was shorter. Eleven of thirteen patients were satisfied with the result, and two patients experienced pain associated with glenoid loosening and persistent anterior subluxation. It was mentioned that most of the complications occurred during the early stages of the clinical trial, indicating that experience of the surgeon was important to the success of the implant.

Failures 8%, Complications 23% (instability, glenoid loosening-related complications).

Amstutz, Sew Hoy and Clark, (1981) 11 implants, 30 month average follow-up:
This early study on the UCLA developed, Dana system showed 9% failures, with a 27% complications rate. Significant improvement in the range of abduction was reported in all 11 patients. Muscle deficiency limited the improvement for one of the patients while traumatic shotgun damage limited the prognosis for another. The postoperative evaluations for pain, function and range of motion improved for both of these patients, however, the shotgun blast victim required a modified glenoid component for more constraint. For the remaining cases, the only other problem stemmed from impingement due to inadequate rotator cuff. Revision, with a hooded glenoid component, was planned at the time of publication.

Failures 9%, Complications 27% (instability related).

Cofield, (1979) 50 implants, 17 months average follow-up:
This trial occurred early in the development of the Neer II prosthesis, and unconstrained shoulder arthroplasty in general. Two complications occurred when the surgeon damaged axillary nerves. Thirty of the 50 shoulders showed cuff inadequacy with 6 being large tears, 92% of the shoulders had mild or no postoperative pain. Glenoid loosening was the probable cause of pain in one of the four shoulders with moderate postoperative pain. Two were due to continuing periarthritis, and one was unexplained. In 22% of the shoulders 90% of active abduction was not possible postoperatively. Of these, eight were rotator cuff deficient and three had tuberosity osteotomy nonunion. The vast majority of the shoulders regained enough function to perform the basic daily functions of dressing, eating unaided and tending to personal hygiene and toilet needs (at least 96%). All patients felt they were improved by the operation, with 72% saying they were much better.

Failures and complications rates could not be deduced from this paper.

Semi-constrained total shoulder arthroplasty

(a) System description

Rotator cuff damage was recognized early in clinical evaluation of unconstrained shoulder replacements as a major limitation. Laurence (1991) developed a system in which the humeral end is loosely constrained by a glenoid component which envelopes the humeral head but is not a socket. The range of motion is not fully constrained because the ball can 'pull' in an outward direction and still be constrained in the lateral directions. Thus, subluxation is reduced. The materials used for this system are stainless steel for the humeral component and UHMWPE. Both components are cemented with PMMA, with reinforcement from screws in the glenoid component. Unlike the Neer and Dana systems, the glenoid component is screwed into the scapula, and it is fixed to all three contact points of the scapula, the acromion, the coracod and the lower lip of the glenoid. A typical, semi-constrained prosthesis is illustrated in Laurence (1991).

(b) Clinical results

Laurence, (1991) 71 implants, 82 months average follow-up:
Only one trial of this type of implant was studied. It was, however, the longest in duration of any examined. There was one case of humeral loosening, with no apparent explanation, and one neuropathic joint dislocated, with subsequent loosening. One other shoulder became ankylosed with fibrous tissue and remained completely stiff, even following surgical clearance, two years after the initial arthroplasty. The remaining complications were loosening due to fall-induced fractures and a dislocation due to hyperextension of the shoulder while lifting. Overall, 75% of the shoulders yielded more than 90 degrees of active abduction after surgery, 93% experienced reduced pain, movement improved in 97% and 94% gained in functional capacity.

Failures 3%, Complications 7% (4% were accident related).

Constrained total shoulder arthroplasty

(a) System description

Initial attempts at shoulder replacements focused on the development of the constrained, or fixed-fulcrum type, in which the 'ball' of the humeral component articulates within a socketed glenoid component, which is also fastened to the scapula. These components have progressed to a state where they are a viable alternative in nearly all replacement surgeries, especially those with a need for joint stability due to soft-tissue deficiency. However, these implants have a very limited resultant range of motion (ROM) and kinematic compromise in which large loads can cause loosening and failure.

Generally, a metal, humeral stem/ball component is attached to a metal encased UHMWPE socket, which is in turn cemented or screwed to the scapula. The attachment of the stem to the humerus is achieved through cementing, bone ingrowth, pressfit or morphological fixation.

The selection of the proper length of humeral component, and the alignment of the implant in the body are critical in keeping the loading of the implant/bone interface to a minimum. The advantages of the constrained type of prosthesis depend on the stability of the resulting joint. They resist the tendency for dislocations and subluxations and provide a substitute for the stability which is compromised in cases of deficient rotator cuff function.

The history of constrained total shoulder arthroplasty has its origins in 1893 (Lugli, 1978). The primary use of constrained total shoulder implants is now for shoulders with serious soft-tissue compromise or a similarly debilitating condition. It is generally accepted that unconstrained implants are superior in most cases where soft tissue condition allows them. Since they are now used primarily in 'problem' shoulders, it is to be expected that they have lower success rates than the other systems. The clinical study reported below addresses constrained prostheses with these considerations in mind.

(b) Clinical results

Post, (1983) 78 implants, 46.5 months average follow-up:
The prosthesis examined in this study had a standard humeral component constrained by a split UHMWPE socket, assembled from halves, inserted into a mating metal component and then secured by a tightened metal ring. Two series of Michael Reese designs were evaluated. Since Series I

prostheses were not implanted after the improved Series II prosthesis was developed, only the performance of the Series II design is evaluated here.

Osteoarthritis and capsular insufficiency were the chief indications for use of this implant. Intractable pain was the chief reason for implant replacement. None of the shoulders had attached rotator cuffs, so it was not expected that the patients would regain much active function, with the exception of those possessing very healthy deltoid muscles. Among the 74 patients, there had been 41 previous procedures before undergoing constrained arthroplasty. The author stressed the importance of recipient screening in use of the constrained prosthesis. One recurring problem patient had good shoulder characteristics, but as a result of alcoholism, frequently dislocated the repaired shoulder, necessitating multiple revisions.

The two failures which were not due to trauma were caused by broken screws. The author concluded that cast screws were not strong enough; cold-worked screws were prescribed for subsequent implants.

Failures 3%, Complications 21% (trauma and screw breakage).

Unconstrained shoulder hemi-arthroplasty

(a) System description

The fourth subset of shoulder implant types is that of hemi-arthroplasties. In these implants, no glenoid component is used. The existing glenoid bone and associated tissue must be adequate to withstand the bearing of the loads of the smooth humeral implant surface. The complications involving this type of system obviously do not involve failure of a glenoid component, thus reducing the potential for implant failure. However, if the existing glenoid surface and the supporting musculature is inadequate, shoulder instability and tissue trauma can result.

A unique variation on hemi-arthroplasty shoulder implants is re-surfacing with a metal cup cemented over the proximal end of the humerus, providing a new, smooth articulating surface. This is the least invasive of all available arthroplasties, preserving the medullary cavity and reducing the overall trauma experienced by the shoulder.

(b) Clinical results

Kay and Amstutz, (1988) 15 implants, 33 month average follow up:
This study evaluated the success of both Dana and Neer humeral components. Both press-fit and cement attachment methods were used. It was specifically stated that neither age nor use of cement affected surgical outcome. One might infer that age plays a lesser role in hemi-arthroplasty since the condition of the scapular bone is not associated with failure or success. This trial population is however too small to draw any general conclusions.

One interoperative humeral penetration occurred, as did one successfully healed infection. One dislocation occurred in the immediate postoperative period, underwent closed reduction and five and one-half years later had experienced no recurrence. One last complication was a torn rotator cuff, which was repaired and at the time of the report, was asymptomatic. Overall, the results of this study were promising, with good pain reduction and fairly good postoperative range of motion and activity levels.

Failures 0, Complications 13% (humeral cortex broached, dislocation with successful closed reduction).

Jonsson, *et al.*, (1986) 26 implants, 24 months average follow-up:
This study evaluates a unique and evidently, highly effective system where the humeral head is prepared and fitted with a stainless steel cup, which replaces the articulating surface. The application was exclusively for sufferers from rheumatoid arthritis. Postoperative pain was not experienced by any patients, and only 4% experienced reduced mobility. All but one post-operative complication cleared-up spontaneously, that one was successfully corrected through revision to resect the acromion, eliminating impingement.

Failures 0, Complications 4% (acromion impingement).

Sim, *et al.*, (1980) 7 implants. 11.4 months average follow-up:
Another unique system was examined in this study. The implant used by Sim was a three-piece unit consisting of an extracortical sleeve, a spacer and a head. All three were made of alumina ceramic. The sleeve is press-fit over the humeral stump with a screw inserted to prevent rotation and to provide initial clinical stability. This system was used in patients with severe damage to the proximal humerus after tumor removal, making the placement of a intermedullary stem unlikely. Extensive resection, accompanied by severe

soft tissue damage, was required. Pain was eliminated in all cases, although active range of motion was minimal, due to the lack of muscle tissue.

This system appears to be a good choice for patients with little chance of regaining active control of the shoulder, because pain relief is good and resulting passive range of motion is excellent. The length and magnitude of this study was not enough to evaluate long-term issues such as endurance of the ceramic sleeve and bone resorption due to stress-shielding at the bone-implant interface. The clinical state of the patient and medication for the underlying condition could be significant factors in long term survival.

Failures 0, Complications 43% (1 dislocation requiring shortened implant component, 1 fractured humerus and 1 spontaneously-healed nerve palsy).

Bipolar implant shoulder arthroplasty

(a) System description

This system was designed primarily as a salvage procedure for arthritic shoulders with torn rotator cuffs. It also has an application where a Neer-type implant cannot be placed without glenoid bone-grafting. The bipolar implant configuration, (see Swanson, 1989 for an illustration) is in a sense, a hemi-arthroplasty implant. The implant is inserted within the humerus in a fashion similar to the other implant configurations. However, no glenoid component is attached to the scapula. Instead, a ball with a socket within it, is placed in the natural socket of the shoulder and may articulate to some extent in the socket. Within this ball is the socket that constrains the proximal ball-end of the humeral component. This pair of nested, ball-and-socket joints provides an excellent range of motion when properly positioned in a relatively healthy shoulder. When subluxation or minor dislocation occurs the joint still should maintain a limited range of motion through motion between the inner cup and the head of the prosthesis.

(b) Clinical results

Swanson, *et al.*, (1989) 35 implants, 63 month average follow up:
The bipolar implant examined in this study was designed in 1975 by the senior author. Twenty one (89%) of the shoulders experienced good to excellent pain relief, 2 patients (6%) were restricted to light activities due to

pain. Improvement in activities of daily living and range of motion were also recorded for every member of the group. The proximal end of the implant tends to sublux in the superior direction progressively, postoperatively. All postoperative complications were correctable and caused no adverse effects upon the performance of the implant.

Failures 0, Complications 14%.

Lee and Niemann, (1994) 14 implants, 40 month average follow-up: This study evaluates the bipolar implant in applications where every shoulder had either massive (more than 5 cm) rotator cuff tears or a thinned, scarred, incompetent rotator cuff. Seven patients had, as their primary indication, rheumatoid arthritis and seven had failed reconstructive procedures. Those with failed prior reconstruction averaged 2.1 previous procedures per patient.

The arthritic shoulders showed significantly reduced pain after follow-up while the reconstructive group saw improvement in three of the four shoulders. Range of motion results again favored the arthritic group. Most normal activities were difficult or impossible for the reconstructive group while the arthritic patients were able to accomplish most tasks without assistance. The UCLA ratings were 36% lower, on average, for the reconstructive group than for the arthritic group.

The only complication in the arthritic group was a fractured humerus, between the distal stem of the humeral prosthesis and the proximal end of a total elbow prosthesis. Six complications occurred in the other group. Dislocation and subluxation plagued 3 patients with no acceptable outcome. The others experienced complications related to tissue trauma from previous procedures.

It is clear from this study that this prosthesis performed adequately, if not well, in all of the arthritic patients with dysfunctional rotator cuffs, and remains an option for the orthopaedic surgeon to deal with these indications. It is equally clear that this is not a solution for other failed repair procedures of the shoulder.

Failures 29%, Complications 50% (all with previous repair procedures).

CONCLUSIONS

The overall failure and complication rates are shown in Fig. 7.2, for the system types examined. These results are combined figures from a variety

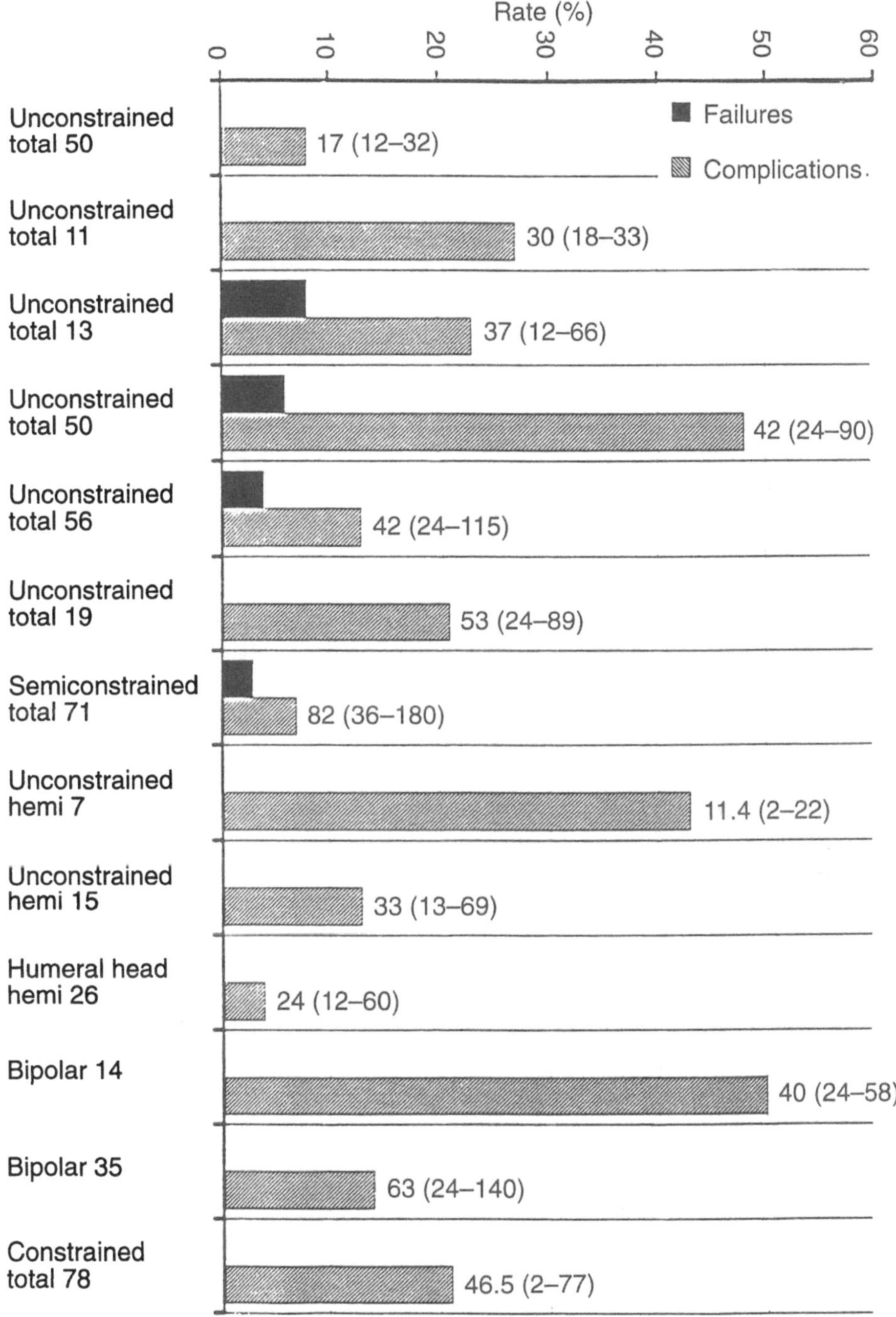

Fig. 7.2 Overall failure and complication rates of shoulder prostheses: values are mean time in months with range in parentheses.

Table 7.5 Summary of shoulder prostheses performance

Reference	System type	Humeral	Glenoid	Follow-up period (months)	Quantity & indications	Rating method
Neer and David 1988	Unconstrained total w/glenoid bone grafting (severe glenoid deficiency)	Neer II	UHMWPE standard metal-backed 200% metal-back PMMA	53 (24-89)	6Rh 5 TrAr 2 Os 3 D 1 CufTr 1 OErb 1 Rev 1 Tr 19 Total	Subjective
Amstutz *et al.* 1988	Unconstrained total	Dana Vitallium alloy PMMA	UHMWPE PMMA 10 hooded 46 regular	42 (24-115)	24 Os 18 Rh 8 TrAr 3 ON 3 Rev 56 Total	UCLA rating
Barrett *et al.* 1987	Unconstrained total	Neer II Co-Cr 36 PMMA 14 Press-fit	UHMWPE PMMA	42 (24-90)	33 Os 11 Rh 6 Fr 50 Total	Am. Elbow & Should. Surgeons eval. form
McElwain and English 1987	Unconstrained total	Author's porous coated	Metal-backed, porous coated w/ UHMWPE liner. Pin & screw & tissue attach.	37 (12-66)	6 Tr-or-ED 5 Rh 1 VS 1 Ps 13 total	UCLA rating
Amstutz *et al.* 1981	Unconstrained total	UCLA-TSA (Dana) Co-Cr-Mo PMMA	UHMWPE PMMA	30 (18-33)	3 TrAr 4 Os 2 Rh 1 ON 1 Tr 11 Total	UCLA rating
Cofield 1979	Unconstrained total	Neer Cr-Co press-fit 46 PMMA 4	UHMWPE PMMA	17 (12-32)	18 Rh 17 Os 15 TrAr 50 Total	Subjective

Indications Legend:

AcFr	Acute 4-part prox. humerus fracture	D	Dislocation
Ar	Unspecified arthritis	De	Degenerative arthritis
AvN	Avascular humeral head necrosis	ED	Epiphyseal dysplasia
Cap Ins	Capsular insuffiency	Fr	Old fracture
CufTr	Cuff tear arthropothy	GCT	Giant cell tumor
ChSar	Chondrosarcoma	Np	Neuropathic arthritis
		OErb	Old ERB brachial-plexus palsy

Results (Pre/Post)				No. of failures	No. of complications	Reference
Pain	ROM	Acitvity	Overall			
			16 excellent 1 satisfact. 2 lim. goals	0	4(21%)	Neer and David 1988
3/8 2.8/7.8 4.9/8.3 3/9 3.3/7 3/8	3/7 2.8/5.8 3.9/5.5 3/9 3.7/5.7 3/6.8			2 (4%)	7 (13%)	Amstutz *et al.* 1988
Pain moderate or greater (pre/post) 48/6	Daily act pre/post 16/82 15/85 10/50 15/79		Results of self-evaluation: 94%-much better off	1 (3%) 2 (33%) 3 (6%)	24 (48%)	Barrett *et al.* 1987
Satisfied in 11/13 (85%)	Improved in 7/13 (54%)	Improved in 11/13 (85%)		1 (8%)	3 (23%)	McElwain and English 1987
4.3/10 2.5/9.8 2.5/8.5 3/10 2/2 3/8.3	2.3/6.3 3.3/7.8 3.5/5.5 3/8 2/4 2.9/6.5	3.3/6.7 4.8/7 5/5 4/8 2/4 4.1/6.4	9.9/23 10.6/24.6 11/19 10/26 6/10 10/21.2	0	3 (27%)	Amstutz *et al.* 1981
100% pre-mod/sever 8% post-moderate	100% dress 78% comb 96% utensil etc. 96% perineal 87% ok sleep		36 much-better 14 better	0	4 (8%)	Cofield 1979

ON	Osteonecrosis of humeral head	Rev	Revision
Os	Osteoarthritis	Rh	Rheumatoid arthritis
Os Sar	Osteosarcoma	Se	Septic arthritis
Po-Tr	Post-traumatic involvement	Tr	Trauma
Ps	Psoriatic arthritis	TrAr	Traumatic arthritis
Rec	Reconstruction of prior surgery	Tu	Resectable tumor of prox. humerus
		VS	Villonodular synovitus

Table 7.5 (continued)

Reference	System type	Humeral	Glenoid	Follow-up period (months)	Quantity & indications	Rating method
Lawrence 1988	Semi-constrained (snap-fit) total (deficient rotator cuff)	Standrd st. st. humeral comp England	UHMWPE PMMA/screws France	82 (36-180)	49 Rh 10 De 7 TrAr 3 AvN 1 Se 1 Np 71 Total	Scale 0-3 0=bad 3=excel.
Kay and Amstutz 1988	Unconstrained hemiarthroplasty	UCLA Dana & Neer w/ 7 press-fit & 8 PMMA	None	33 (13-69)	11 AcFr 3 AvN 1 Os 15 Total	UCLA rating
Sim *et al.* 1980	Unconstrained hemiarthroplasty (proximal hem. tumor resect.)	Ceramic (Al_20_3) conical fit and tissue ingrowth	None	11.4 (2-22)	2 Chsa 2 OsCa 3 GCT 7 Total	Subjective
Jonsson *et al.* 1986	Humeral head cup	St. st. cup Palacos cem in gentamicin	None	24 (12-60)	26 Rh 26 Total	Measured ROM, rel. pain
Swanson *et al.* 1989	Bipolar	Titanium PMMA	Co-alloy shell UHMWPE liner	63 (24-140)	20 Rh 10 Os/De 5 Po-Tr 35 Total	0-10 bad-good 3 catagor.
Lee and Niemann 1994	Bipolar rotator cuff tear - salvage oper.)	Titanium PMMA	Co-Cr shell UHMWPE	40 (24-58)	7 Rh 7 Rec 14 Total	UCLA and relative
Post and Jablon 1983	Constrained total	Co-Cr PMMA	Split UHMWPE socket in metal glenoid PMMA w/screws	46.5 (2-77)	36 Ar 8 Rev 10 ON/Fr 2 Tu 8 D or Fr/D 14 Cap Ins 78 total	Arbitrary

Results(Pre/Post)				No. of failures	No. of complications	Reference
Pain	ROM	Activity	Overall			
0/2	0.1/1.4	0/1.7		1 2 (3%)	1 5 (7%)	Lawrence 1988
7.2 2.7/9.3 8 7.6	5.5 8 5 6	6.9 8.7 6 7.2	19.6 26 19 20.8	0	2 (13%)	Kay and Amstutz 1988
no pain in all cases	excellent passive ROM	minimal active ROM	implant acts simply as a spacer	0	3 (43%)	Sim *et al.* 1980
NO Pain 100%	4% reduc. mobility 21% no chg 75% improv.			0	1 (4%)	Jonsson *et al.* 1986
2/9 2/8 3/9 2/9	6/8.2 5.6/7.8 6.8/8.8 6/8	4/7 4/7 6/9 4/7	3.3/7.4 3.9/7.6 5.3/8.9 4/8	0	5 (14%)	Swanson *et al.* 1989
1-Gd:5-Bd 4.7/2 4.9/3.4		1-Bd:5-Gd 3.6/3.7 3.3/3.2	UCLA 6.7/17.1 5.3/11	0	1 6 7 (50%)	Lee and Neimann 1994
20 Ex 30 Gd 17 Fr 3 Pr	60% improved active fnctn 65% improved passive fnctn			0	16 (21%)	Post and Jablon 1983

of specific studies, and only indicate general trends. For instance, even though unconstrained total arthroplasties show more failures than any other system, the experience level and the length of the clinical trials exceeds that of nearly every other system. Similarly, the apparent superiority of the humeral head cup is based on a clinical trial on a small number of shoulders, and should be more closely compared with the other systems to determine the reasons for such success.

Table 7.5 is a summary of the shoulder implant systems evaluated in this chapter. Figure 7.3 shows the comparative failure and complication rates of these systems. From these data, it is clear that the condition of the shoulder and the patient's history are the determining factors for success of a particular implant system.

For the systems examined, it appeared that the use of cement did not affect the failure rate of any system, thus it is recommended that it be used in both humeral and glenoid fixation. Screws should be used in the glenoid component of constrained implants and should be the maximum strength, surgical quality available.

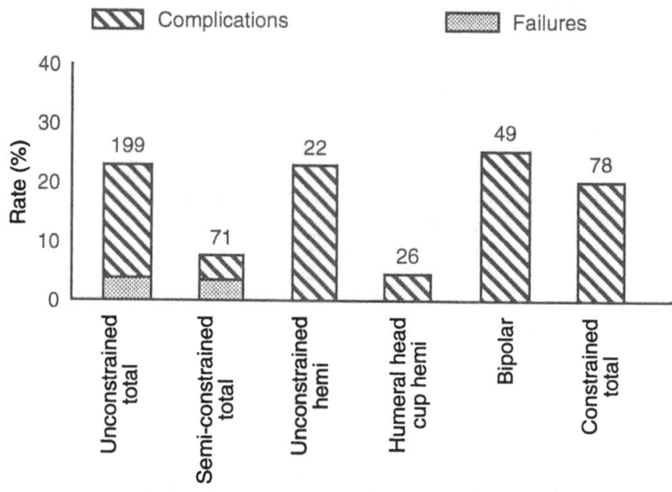

Fig. 7.3 Failure complication rates for various shoulder prostheses systems.

In general, in the absence of contraindications, the order of preference for the selection of a shoulder prosthesis is,

1) none
2) unconstrained hemi-arthroplasty
3) unconstrained total shoulder arthroplasty
4) semi-constrained total shoulder arthroplasty
5) constrained total shoulder arthroplasty

The unconstrained hemi-arthroplasty yields a greater postoperative range of motion and function, as well as satisfactory pain reduction, with proper operative procedures and postoperative therapy. The absence of the glenoid component gives fewer parts to fail and a less complicated surgical procedure, which reduces the chance of surgical error.

In the case of severe rotator cuff deficiency, the semi-constrained total shoulder arthroplasty appears to have performed the best. This trial included a sizable population and up to ten years follow-up. The failure rate was unexpectedly low, considering the condition of many of the shoulders. The bipolar implants seemed to have unpredictable results, depending on the patient and the geometry of the sub-acromional arch. This makes it difficult to define appropriate contraindications, making success of this system very dependent on the ability and judgement of the surgeon.

Of all of the systems examined, the most promising system appears to be the humeral cup hemi-arthroplasty. Its success rates were as high as any other, with fewer complications.

REFERENCES

Amstutz, H.C., Sew Hoy, A.L. and Clarke, I.C. (1981) UCLA anatomic total shoulder arthroplasty. *Clinical Orthopaedics and Related Research* **155,** 7-20.

Amstutz, H.C. *et al.*, (1988) The Dana total shoulder arthroplasty. *The Journal of Bone and Joint Surgery* 70-A, 1174-81.

Barrett, W.P. *et al.,* (1987) Total shoulder arthroplasty. *Journal of Bone and Joint Surgery* **69-A**, 865-72.

Cofield, R.H. (1979) Total joint arthroplasty, the shoulder. *Mayo Clinic Proceedings* **54,** 500-6.

Jonsson, E. *et al.,* (1986) Cup arthroplasty of the rheumatoid shoulder. *Acta Orthopedic Scandinavica* **57**, 542-46.

Kay, S.P. and Amstutz, H.C. (1988) Shoulder hemiarthroplasty at UCLA *Clinical Orthopaedics and Related Research* **228,** 42-8.

Lawrence, M. (1991) Replacement arthroplasty of the rotator cuff deficient shoulder. *Journal of Bone and Joint Surgery* **73-B,** 916-19.

Lee, D.H. and Niemann, K.M.W. (1994) Bipolar shoulder arthoplasty, *Clinical Orthopaedics and Related Research* **304,** 97-107.

Lugli, T. (1978) Artificial shoulder joint by Pean, *Clinical Orthopaedics and Related Research* **133,** 215-18.

McElwain, M.B. and English, E. (1987) The early results of a porous-coated total shoulder arthroplasty, *Clinical Orthopaedics and Related Research* **218**, 217-24.

Neer, C.S. II (1985) Unconstrained shoulder arthroplasty, *Instructional Course Lectures* **34**, 278-86.

Neer, C. S. II and David S. (1988) Glenoid bone-grafting in total shoulder arthroplasty, *Journal of Bone and Joint Surgery* **70-A**, 1154-62.

Post, M. and Jablon, M. (1983) Constrained total shoulder arthroplasty, *Clinical Orthopaedics and Related Research* **173**, 109-16.

Sim, F. H. *et al.* (1980) Replacement of the proximal humerus with a ceramic prosthesis: A preliminary report. *Clinical Orthopaedics and Related Research* **146**, 161-74

Swanson, A. B. *et al.* (1989) Bipolar implant shoulder arthroplasty, long-term results, *Clinical Orthopaedics and Related Research* **249**, 227-47.

8

Elbow Joint Implant Systems

John Delvaux

INTRODUCTION

The goal of any arthroplasty is to restore and maintain mobility and to eliminate pain. The three major functions of the elbow joint are: (1) to serve as a link in the upper extremity system that positions the hand, (2) to provide a fulcrum that allows the elbow flexors to generate sufficient force for lifting, and (3) to serve as a weight bearing joint in individuals requiring walking aids. All currently used prostheses provide satisfactory motion for of daily living (Bryan and Morrey, 1979; Inglis and Pellicci, 1980).

About 70% of the patients who are candidates for total elbow arthroplasty have rheumatoid arthritis. Those with post-traumatic arthritis account for about 20% and the remainder is for various other reasons. If good bone stock and stability are present before surgery then the goals can be met by simply resurfacing the articulating joint. This has been done in a limited number of studies with silicone polymer as a radial head replacement. If instability or pronounced bone loss exists the goals can be achieved by a semiconstrained prosthesis. Indications for each of these procedures depend on the spectrum of pain, loss of motion, and instability (Inglis and Pellicci, 1980).

Studies have shown that rheumatoid arthritis patients fare better with total elbow replacement than do post-traumatic arthritis patients. The patient with rheumatoid arthritis is usually unable to exceed the limitations of the prosthesis or its fixation. Conversely, the patient with post-traumatic arthritis is more active and expects more from the joint. The procedure more often fails in this type of person (Bryan and Morrey, 1979).

Design considerations of any joint must include kinematics. The elbow joint kinematics are complex with 180 degrees between supination and pronation of the hand and 150 degrees between extension and flexion (Fig. 8.1). During lifting, a force of approximately 2.6 times the weight of the

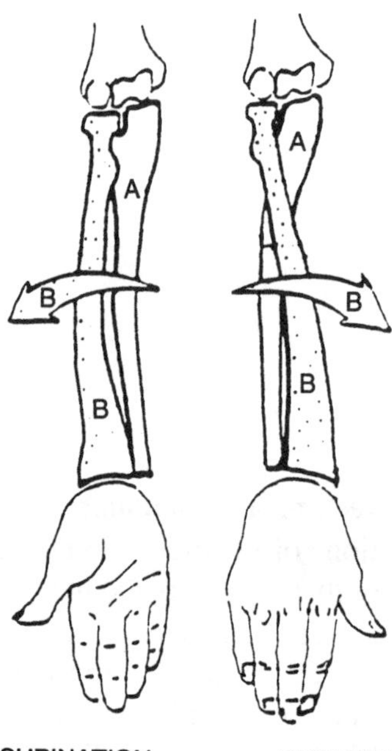

Fig. 8.1 Elbow joint kinematics.

body is transmitted across the elbow joint (Bryan and Morrey, 1979). Others have measured forces 5 to 6 times body weight in any direction of motion, with varying degrees of torque (Davis *et al.*, 1982). With elbow flexion, the forces are cyclically directed anteriorly and posteriorly at the distal humerus. In the 1970s elbow joints were fully constrained, that is pinned with one degree of freedom. This proved to be too restrictive for the large and complex loads across the elbow, resulting in failures, mostly by loosening. Therefore, constrained hinged arthroplasty was abandoned. By 1980 semi-constrained and unconstrained designs were developed with the intention of unloading the bone-cemented interface by dispersing the forces into the soft tissue (Rosenberg and Turner, 1984).

Because of the special elbow kinematics discussed above, several design principles have been proposed for joint type elbow prostheses. Firstly, the simple hinge motion does not replicate the normal elbow motion, and hence a constrained prosthesis should have some play at the hinge axis. Secondly, the distal humeral component undergoes cyclic loading and unless the prosthesis provides good fixation at the distal humerus there is a tendency to loosen. Thirdly, prostheses designed to allow radiohumeral articulation can distribute loads more uniformly and may provide better strength by stabilizing the insertion of the biceps on the proximal radius. Fourthly, the prosthesis should provide a low friction joint. Finally, the procedure should require the resectioning of as little bone as possible so that a soft-tissue arthroplasty could still be performed as a salvage procedure in the event of failure (Bryan and Morrey, 1979).

DISCUSSION

There are 8 different types of total elbow joint arthroplasty systems that have been extensively tested; (1) GSB III, (2) Capitelloncondylar, (3) Coonrad, (4) HSS-Osteonics,® (5) Mayo, (6) Pritchard-Walker, (7) Triaxial-prosthesis, and (8) Type 1&2. There was one partial elbow joint arthroplasty system also extensively tested, the Silastic® radial head. These systems were reviewed and the findings are discussed. Clinical data were gathered on all of these systems and are shown in Table 8.1. The systems are shown in alphabetical order, the date reflects completion of the study. The materials of each system are shown, but there were several variations on their use, discussed below. Polyethylene and cobalt-chrome were used in nearly every type. Polymethylmethacrylate cementing was the fixation method used in all systems. The cause of the natural joint failure is shown under indications. The average time between surgery and follow-up us shown under period. The number of elbow implants followed-up is shown. The preoperative and postoperative score shown gives a quantitative value to the quality of life before and after surgery. Details on scoring are given below. The number of implant failures is a subset of the number of complications. Failures consisted of implant component failure, implant loosening, implant dislocation and implant disassembling. There were a variety of other complications, listed in Table 8.2.

Table 8.1 Summary table for elbow implant study

Reference	System	Material	Indications
Gschwend *et al.* 1988	GSB III	Metal/Polyethylene	rheumatoid/traumatic arthritis, other
Weiland *et al.* 1989	Capitellon.-Kocher	Cobalt–chrome/Polyethylene	rheumatoid arthritis, other
Ewald *et al.* 1980	Capitelloncondylar	Cobalt–chrome/Polyethylene	rheumatoid arthritis
Davis *et al.* 1982	Capitelloncondylar	Cobalt–chrome/Polyethylene	rheumatoid/osteoarthritis arthritis
Ewald *et al.* 1993	Capitelloncondylar	Cobalt–chrome/Polyethylene	rheumatoid arthritis
Rosenberg 1984	Capitelloncondylar	Cobalt–chrome/Polyethylene	rheumatoid arthritis
Trancik *et al.* 1987	Capitelloncondylar	Cobalt–chrome/Polyethylene	rheumatoid arthritis
Bryan 1977	Coonrad	Metal/Polyethylene	rheumatoid arthritis, other
Bryan 1979	Coonrad	Metal/Polyethylene	rheumatoid/traumatic arthritis, other
Bryan 1981	Coonrad	Metal/Polethylene	rheumatoid arthritis
Bryan 1981	Coonrad	Metal/Polyethylene	post-traumatic arthritis
Morrey and Adams 1992	Modified Coonrad	Metal/Polyethylene/Porous	rheumatoid arthritis
Kraay *et al.* 1994	HSS–Osteonics	Metal/Polyethylene	rheumatoid arthritis
Kraay *et al.* 1994	HSS–Osteonics	Metal/Polyethylene	post-traumatic arthritis
Bryan 1977	Mayo	Metal/Polyethylene	rheumatoid arthritis, other
Bryan 1979	Mayo	Metal/Polyethylene	rheumatoid/traumatic arthritis, other
Bryan 1981	Mayo	Stainless-Steel/Polyethylene	rheumatoid arthritis
Bryan 1981	Mayo	Stainless-Steel/Polyethylene	post-traumatic arthritis
Inglis and Pellicci 1980	Pritchard–Walker	Cobalt–chrome/Polyethylene	rheumatoid arthritis
Inglis and Pellicci 1980	Pritchard–Walker	Cobalt–chrome/Polyethylene	post-traumatic arthritis
Morrey *et al.* 1981	Silastic radial head	Silicone	unknown
Mackay *et al.* 1982	Silastic radial head	Silicone	rheumatoid arthritis
Ferlic *et al.* 1987	Silastic radial head	Silicone	rheumatoid arthritis
Inglis and Pellicci 1980	Triaxial-Prosthesis	Cobalt–chrome/Polyethylene	rheumatoid arthritis
Inglis and Pellicci 1980	Triaxial-Prosthesis	Cobalt–chrome/Polyethylene	post-traumatic arthritis
Kudo and Iwano 1990	Type 1&2	Metal/Polyethylene	rheumatoid arthritis

n/a = not available.

Period years	Implants	Pre-op Score	Post-op Score	Failures	Complications
4.1	63	31	91	7	19
7.2	30	30	88	2	11
3.5	54	36	90	1	21
3.3	30	8	85	0	9
2	54	n/a	90	4	8
2.9	28	29	89	6	14
5.6	31	n/a	97	4	20
3	26	n/a	50	12	n/a
4	35	n/a	84	5	7
4.1	15	n/a	85	7	15
4.1	18	n/a	72	4	11
3.8	58	38	94	0	15
5	86	n/a	n/a	9	n/a
5	6	n/a	n/a	3	n/a
2	26	n/a	71	2	n/a
4	50	n/a	84	8	10
4.1	44	n/a	81	6	17
4.1	3	n/a	56	0	1
3.7	11	33	82	1	7
3.7	4	40	46	1	2
6.7	17	n/a	n/a	8	8
2	18	n/a	n/a	0	0
7.2	57	n/a	78	0	8
3.7	11	33	82	1	8
3.7	5	40	46	1	2
9.5	17	44	93	2	5

Table 8.2 Complications of elbow prostheses

Permanent	*Transient*
Infection	Dislocations
Disassembling	Ulnar nerve paresis
Loose prosthesis	Epicondylar fracture
Fractured prosthesis	Ulnar fracture
	Ulnar perforation
	Synostosis
	Superficial skin problems
	Triceps rupture
	Delayed wound healings
	Hematoma
	Humerus fracture

Score

About half of the studies used an Ewald scoring system on preoperative and postopertative evaluations. The test rates pain, function, range of motion, and deformity on a cumulative 100 point scale. Pain is assigned 50 points, function 30 points, motion 10 points, flexion contracture 5 points, and deformity 5 points. An excellent score is 90 to 100, a good score, 80 to 89, a fair score, 70 to 79, a poor score, 69 or less. Many other studies used a modified version of the Ewald system with greater emphasis on degrees of pain, motion and stability. A few studies relied on a subjective view of pain and function. In all studies relief from pain was the most significant indicator of success. The scoring technique is not consistent for all studies, but there were sufficient similarities to use the score value for comparison.

Types of devices available

Illustrations of all these types may be found in the appropriate references.

(a) GSB III

Today semiconstrained or condylar prostheses are the options when replacing destroyed elbow joints. These are the new generation of elbow prostheses, but there is still a high complication rate in both types. The GBS III prostheses is a semiconstrained prosthesis with a ulnar stud and humeral stud for fixing. Both are made of metal which is not specified in the reference, as well as the hinge joint. The ulnar stud glides in a high density polyethylene bushing connected to the hinge. This provides good joint kinematics. A wide condylar flange on the humeral stud allows for good load transmission with minimum bone resection (Gschwend *et al.*, 1988).

(b) Capitelloncondylar

This prosthesis consist of a cobalt-chrome bicondylar humeral component with a fixation stem. The ulnar component has a fixation stem (metal or polyethylene) connected to a metal boat that houses the polyethylene articulating surface. The conformed plastic has a central trochlear ridge to guide joint motion, but there is no linkage between the humeral and ulnar component. The two components are designed to resurface the articulating ends of the humerus and ulna so that a minimum of bone is removed during the surgical implantation. This design allows the surrounding soft tissue to absorb large amounts of torque (instead of the implant), but because this design is less stable than that of hinged implants, success in its use requires that the anatomical relationship between the two components be oriented precisely and that the tensions around each component be balanced (Ewald and Jacobs, 1984). Recent studies show that a surgical approach, called the Kocher approach, protects the ulnar nerve well and effectively eliminates permanent ulnar-nerve damage (Weiland *et al.*, 1989).

(c) Coonrad

The Coonrad prosthesis has a long metallic intramedullary humeral stem with distal humeral fixation. The ulna has a shorter metallic stem. The two are connected by a hinge joint with polyethylene bushings. There is no replacement for the radial head. The modified Coonrad prosthesis uses either titanium with a hydroxyapatite plasma spray or a beaded surface to provide a cementless fixation (Morrey and Adams, 1992).

(d) HSS-Osteonics

This implant design is similar to the Coonrad except that the hinge joint is more bulky to provide better load transmission across the polyethylene bearing and into the bone (Kraay *et al.,* 1994).

(e) Mayo

The Mayo implant is similar to the Coonrad except that it uses a radial head component made from polyethylene with an intramedullary stem. The radial head is not connected to the other components. The hinge joint uses a C-shaped encasement on the ulnar component. This fits around a shaft fixed to the humeral stem. The ulnar component was initially made of polyethylene but was later changed to metal (Bryan *et al.*, 1981).

(f) Pritchard-Walker

This implant has a polyethylene humeral component and a cobalt-chrome ulnar component. A snap fit hinge connects the two (Inglis and Pellicci, 1980).

(g) Silastic® radial head

This implant consist of a contoured silicone disc with stem which replaces the radial head joint. There are no connections to the humerus or ulna (Morrey *et al.,* 1981).

(h) Triaxial-Prosthesis

Similar to the Pritchard-Walker prosthesis, this implant design has a cobalt-chrome humeral component and a polyethylene bearing which facilitates surgical implantation (Inglis and Pellicci, 1980).

(i) Type 1&2

This implant is described as an unconstrained metal-to-polyethylene resurfacing prosthesis. No details were given, but the results of the study are shown for comparison (Kudo and Iwano, 1990).

RESULTS

Total elbow arthroplasty has been done over the last 30 years with modest improvements to the implant design. The chief reason for earlier implant failures was the lack of understanding of elbow kinematics, undeveloped surgical techniques and poor patient screening. Much has been learned over the past 30 years. The newer implants more closely reproduce elbow kinematics, better surgical techniques have been defined, and patient screening has improved.

CONCLUSIONS

From these studies three major conclusions can be made.

All elbow implant types provide substantial improvement in the quality of the patients' lives, (see Fig. 8.2).

Pain levels were significantly reduced if not eliminated.

Range of motion increased in all directions, but least improvement was seen in extension.

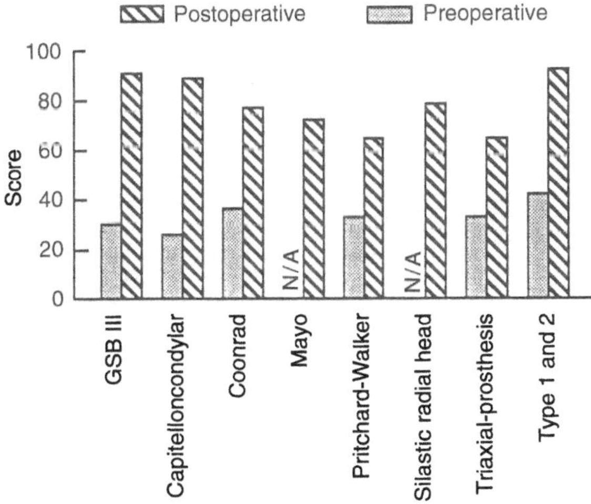

Fig.8.2 Improvement in quality of life with elbow prostheses (Ewald System).

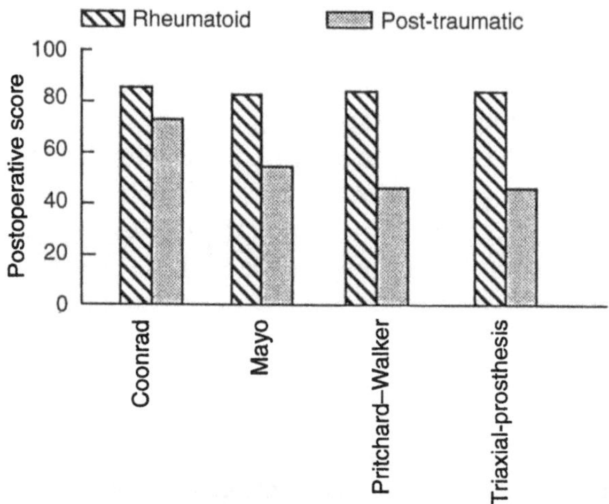

Fig. 8.3 Relative success of elbow prostheses for rheumatoid arthritis patients vs. post-traumatic patients.

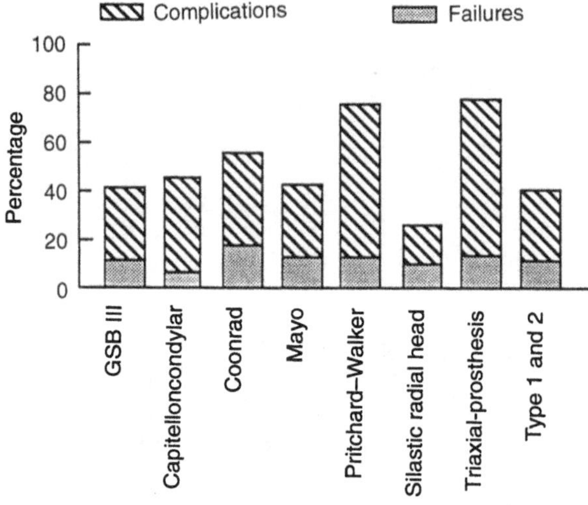

Fig. 8.4 Complication and failure rates of elbow prostheses.

Patients with rheumatoid arthritis improved more with an implant than did post-traumatic patients, (see Fig. 8.3). This is an important screening guideline for new patients. Implant failures were uncommon but the complication rate was high for most implant types (see Fig. 8.4). The Capitelloncondylar system has the lowest failure rate, largely due to the increased degree of freedom of this implant design.

The Coonrad implant failure rate was more than double the Capitelloncondylar, mostly due to the constrained, single degree of freedom, design of the implant.

REFERENCES

Bryan, R.S. (1977) Total replacement of the elbow joint, *Archives of Surgery*, **112** (9), 1092-93.

Bryan, R.S. and Morrey, B.F. (1979) Total joint arthroplasty - the elbow, *Mayo Clinic Proceedings,* **54** (8), 507-12.

Bryan, R.S. and Morrey, B.F., Dobyns, J.H. and Linscheid, R.L. (1981) Total elbow arthroplasty, *Journal of Bone and Joint Surgery,* **63-A** (7), 1O50-63.

Davis, R.F., Weiland, A.J, Hungerford, D.S., *et al.* (1982) Nonconstrained total elbow arthroplasty, *Clinical Orthopaedics and Related Research,* **171** 156-60.

Ewald, F.C. *et al.* (1993) Capitellocondylar total elbow replacement in rheumatoid arthritis, *Journal of Bone and Joint Surgery,* **75A** (4), 498-507.

Ewald, F.C. and Jacobs, M.A. (1984) Total elbow arthroplasty, *Clinical Orthopaedics and Related Research,* **182,** 137-42.

Ewald, F.C., Scheinberg, R.D., Poss, R., *et al.* (1980) Capitellocondylar total elbow arthroplasty, *Journal of Bone and Joint Surgery,* **62-A** (8), 1259-63.

Ferlic, D.C., Patchett, C.E., Clayton, M. L. and Freeman, A.C. (1987) Elbow synovectomy in rheumatoid arthritis, *Clinical Orthopaedics and Related Research,* **220,** 119-25.

Gschwend, N., Loehr, J., Ivoseviv-Radovanonic, D., *et al.* (1988) Semiconstrained elbow prostheses with special reference to the GSB III prosthesis, *Clinical Orthopaedics and Related Research,* **232,** 104-11.

Inglis, A.E. and Pellicci, P.M. (1980) Total elbow replacement, *Journal of Bone and Joint Surgery,* **62A** (8), 1252-58.

Kraay, M. J. *et al.* (1994) Primary semiconstrained total elbow arthroplasty, *Journal of Bone and Joint Surgery,* **76B** (4), 636-40.

Kudo, H. and Iwano, K. (1990) Total elbow arthroplasty with a non-constrained surface-replacement prosthesis in patients who have rheumatoid arthritis, *Journal of Bone and Joint Surgery,* **72A** (3), 335-62.

Mackay, I., FitzGerald, B. and Miller, J.H. (1982) Silastic radial head prosthesis in rheumatoid arthritis, *Acta Orthopedica Scandinavica,* **53** (1), 63-66.

Morrey, B.F., Askew, L. and Chao, E.Y. (1981) Silastic prosthetic replacement for the radial head, *Journal of Bone and Joint Surgery,* **63A** (3), 454-58.

Morrey, B.F. and Adams, R.A. (1992) Semiconstrained arthroplasty for the treatment of rheumatoid arthritis of the elbow, *Journal of Bone and Joint Surgery,* **74A** (4), 479-90.

Rosenberg, G.M. and Turner, R.H. (1984) Nonconstrained total elbow arthroplasty, *Clinical Orthopaedics and Related Research,* **187**, 154-62.

Trancik, T., Wilde, A.H. and Borden, L.S. (1987) Capitellocondylar total elbow arthroplasty, *Clinical Orthopaedics and Related Research,* **223**, 175-80.

Weiland, A.J., Weiss, A.C., Wills, R.P. and Moore, J.R. (1989) Capitellocondylar total elbow replacement, *Journal of Bone and Joint Surgery,* **71A** (2), 217-22.

9

Toe Joint Implant Systems

John Delvaux

INTRODUCTION

Arthritis of the metatarsophalangeal (toe) joint can lead to a painful and debilitating condition. Most toe arthroplasty involves the joint between the metatarsal and phalangeal bones of the big toe, (Fig. 9.1). The indications for toe joint implant arthroplasty are, according to Vanore *et al.* (1984):

1. hallux vagus with subluxation and painful/limited range of motion
2. hallux radius
3. revision surgery
4. rheumatoid arthritis
5. painful degenerative joint disease
6. gouty arthritis
7. osteochondral fractures
8. intra-articular fractures.

There are three types of toe joint implants. One has a silicone rubber disc with a stem which mounts into the metatarsal bone and acts as an articulating surface against the phalangeal bone, (Fig. 9.2). The second type is a silicone rubber hinge with stems at both ends to mount into the phalangeal and metatarsal bones, thereby completely replacing the original joint (Fig. 9.3). The third type uses a stainless steel disc with a polyethylene (PE) stem and mounts into the metatarsal bone. It acts against the phalangeal bone to create an articulating surface.

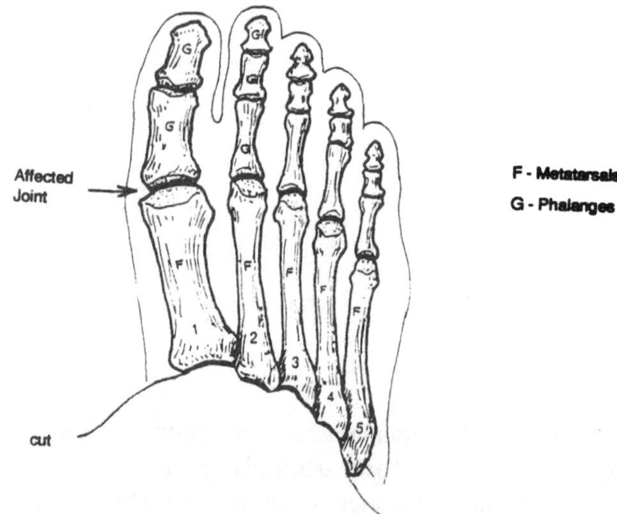

Fig. 9.1 Diagram of foot bones showing the most affected joint.

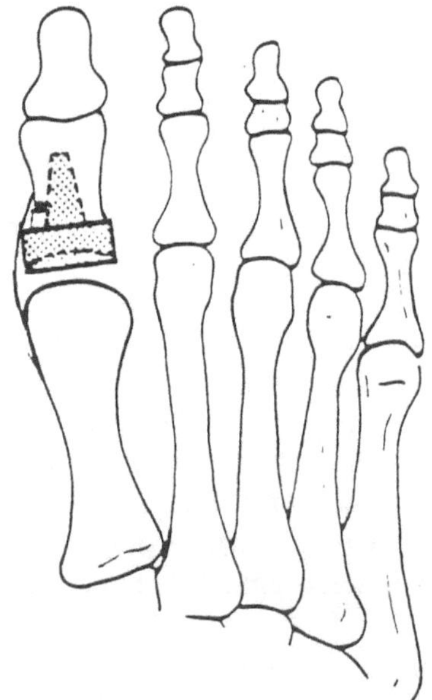

Fig. 9.2 Diagram of a silicone single stem device in place.

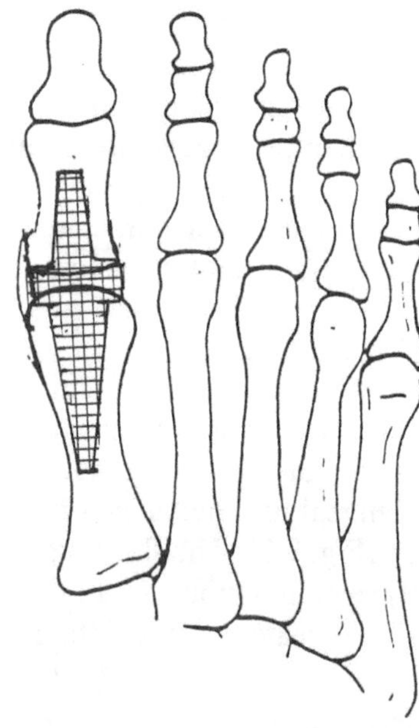

Fig. 9.3 Diagram of silicone hinge type device in place.

DISCUSSION

The first two types listed above are completely made of silicone rubber. These are the most popular types and there have been very few failures. One study followed by Swanson *et al.*, (1979) had 65 single stem implants of which there were only 3 failures and 3 complications. Pain and intolerance to wearing shoes were relieved in most patients. The same study also followed 105 implants with a flexible hinge with no implant failures. Deformities were dramatically improved. However the different follow-up times were unclear. Another study (Verhaar *et al.*, 1989) followed 59 single stem implants, many of which had failed in younger patients. The silicone rubber abraded under load, creating debris around the joint. This caused a bone/soft-tissue reaction characterized by the presence of giant cells around the silicone debris. The mean follow-up time was five years.

Silicone implant failure may be divided into intrinsic and extrinsic causes. Intrinsic failure is the result of the physical properties of the implant material; deformation, fatigue fracture, and microfragmentation. Extrinsic failure is cause by implant tampering or injudicious use. Data for the stainless steel/polyethylene type is limited to only 18 patients (Johnson, 1979). There were two failures but all others experienced the elimination of pain, a reduction in deformity and increase in flexibility. The follow-up time was considered to be too short to be useful.

CONCLUSION

All implant types achieved the desired goals:

1. elimination of pain;
2. reduction in deformity;
3. increase in mobility.

There were a some failures particularly in young, active patients. The implant which had the highest success rate, the silicone hinge with two stems, would be the implant of choice.

REFERENCES

Johnson, K.A. (1979) Total joint arthroplasty, *Mayo Clinic Proceedings,* **54** (9), 576-78.

Swanson, A.B., Lumsden, R.M. and Swanson, G.D. (1979) Silicone implant arthroplasty of the great toe, *Clinical Orthopaedics and Related Research,* **142,** 30-43.

Vanore, J., O'Keefe, R. and Pikscher, I. (1984) Silastic implant arthroplasty, *Journal of the American Podiatry Association,* **74** (9), 423-33.

Verhaar, J., Vermeulen, A., Bulstra, S. and Walenkamp, G. (1989) Bone reaction to silicone metatarsophalangeal joint-1 hemiprosthesis, *Clinical Orthopaedics and Related Research,* **245**, 228-32.

10

Evaluation of Limb Lengthening Techniques

Daniel F. Justin

INTRODUCTION

A major challenge to the orthopaedic research community is to improve the method of bone lengthening so that physicians can offer patients who need it a more effective option than those now available. The first step to improving any surgical method is to understand the advantages and limitations of techniques which have been used clinically. Since Codivilla (1905) first reported surgical lengthening of limbs, significant advances have been made in treatment of deficiencies resulting from trauma, infections, disease, and bilateral length disorders. Current treatment techniques include:

- Equalization by contralateral shortening;
- Orthotic management;
- Direct surgical lengthening;
- Periodic lengthening with an external device followed by bone grafting;
- Gradual osteogenic distraction with an external device;
- Gradual osteogenic distraction with an external device and intramedullary stabilization.

Many difficulties limit the success of each of these methods. Published clinical studies show a wide variety of complications reported with each technique. However, these complications can generally be classified in the following categories:

- Infection;
- Nerve or vascular injury;
- Severe pain;
- Transcutaneous pin loosening;
- Device failure;
- Hip or knee complications;
- Bone fracture after device removal;
- Bone malalignment or nonunion;
- Social and psychological complications;
- Expense.

In this chapter, the indications for lengthening are summarized and the principles of osteogenesis distraction are outlined, the current treatment techniques are described, the associated complications are discussed and the current development of implantable limb lengthening devices is reviewed.

CLINICAL INDICATIONS

The obvious indication for limb lengthening is the presence of a short limb. However, unlike other orthopaedic conditions that clearly require surgical intervention, short limbs do not necessarily need to be surgically lengthened to produce acceptable biomechanical function for the patient. Orthotics are commonly prescribed for patients with less severe discrepancies. Prosthetics are more feasible in cases were lengthening is prohibited or a substantial length deficiency exists.

Indications for lengthening of only one limb usually arise from congenital disorders, trauma, or bone loss as a result of severe infection. Bilateral lengthenings may be for patients with extreme achondroplasia or other dwarfism in which the affliction severely affects the patient's ability to perform daily functions. Many factors influence the choice of treatment. These include the patient's age when the condition is first evident, the growth pattern of the limb, whether the condition is progressive or stabilized, and the patient's psychological and social situation.

Unilateral lengthening may be required after trauma or infection. The severity of open bone injuries is characterized by a rating system developed by Gustilo and Anderson (1976) and Gustilo and Mendoza (1984) and shown in Table 10.1. Open fractures are classified depending on soft tissue damage.

A higher number correlates to a more acute condition and greater likelihood of progression to amputation. A type I fracture corresponds to an open fracture with a clean wound less than 1 cm long. On the other end of the scale, a type IIIB fracture is usually associated with extensive injury to the soft tissue with stripping of the periosteum and exposure of bone, and a type IIIC is associated with arterial injury requiring repair.

As recently as the mid 1980s, attempted reconstruction of grade IIIB and IIIC injuries has resulted in amputation rates as high as 80% (Weiland *et al.*, 1984). Currently, many attempts at reconstruction of type IIIB and greater fractures still end in amputation (Spiro *et al.*, 1993).

One of the main problems in the treatment of severe traumatic injuries and limb infections is the management of extensive soft tissue and bone loss. Multiple secondary grafting procedures are necessary if the initial limb length is to be preserved after such injuries. Before the recent development and refinement of microsurgical tissue transfers, which make well-vascularized soft tissue coverage possible, attempts at limb salvage were jeopardized by the inability to obtain soft tissue coverage of the bone (Spiro *et al.*, 1993). Even with these advances, staged reconstruction can still lead to prolonged hospitalization and a high incidence of complications. The typical staged reconstruction treatment may take two or more years of multiple surgeries and rehabilitation. Severe orthopaedic trauma with open wounds and tissue damage are situations in which infection is almost inevitable (Green, 1983). Allografts and implants cannot be placed into contaminated bone tissues.

Table 10.1 Gustilo and Anderson open fracture classification

Type I.	An open fracture with a clean wound less than one centimeter long without excessive soft-tissue damage, flaps or avulsions.
Type II.	An open fracture with a laceration more than one centimeter long.
Type III A.	Either an open segmental fracture on open fracture with extensive soft tissue damage.
Type III B.	An open fracture with extensive injury to the soft tissue and extensive injury to the periosteum.
Type III C.	An open fracture with excessive soft tissue injury with arterial injury requiring repair.

In cases involving severe trauma or infection, a primary shortening procedure, followed by secondary lengthening has been shown to improve the results of treatment for this difficult problem (Spiro *et al.*, 1993; Betz *et al.*, 1993). The lengthening procedure is usually accomplished by distraction techniques that introduce the risks associated with external fixation.

Limb length inequality may also be congenital or acquired in childhood. Conditions which lead to such discrepancies include poliomyelitis, Ollier's disease, hemiatrophy and Silver's syndrome (Grill and Dungl, 1991; Karger *et al.*, 1993).

Bilateral lengthening of the lower limbs of extremely short statured patients is a rare orthopaedic procedure. Although some attempts to lengthen achondroplastic dwarfs have been moderately successful, the benefits of this elective lengthening seldom outweigh the associated pain and risks (Price 1989a; Aldegheri *et al.*, 1988).

In limb length discrepancies resulting from disease or trauma usually only one bone segment requires lengthening. However, in patients with bilateral length deficiencies associated with achondroplasia and other dwarfism, all the femoral and tibial bone segments are short and can potentially be lengthened. Strategic considerations, such as the order and timing of lengthenings must also be managed such that symmetry is maintained during the process. Thus, if the treatment must be stopped, the limb lengths should remain as close to symmetrical as possible. When more than one limb is lengthened, the lengthening devices used should not interfere with normal biomechanical function during the treatment.

Fewer complications are reported for lengthening procedures on limbs that are short due to disease or dwarfism than for discrepancies following trauma or infection. This is because the lengthening of diseased short limbs does not usually involve extensive soft tissue damage or treatment of the trauma or infection.

Limb length inequality is a major problem worldwide but few reports have documented the number of people with limb length discrepancies. In a French epidemiological study, which surveyed all citizens receiving government assisted medical treatment, the incidence of lower limb length inequality of more than 2 cm was at least 1 in every 1000 persons. The incidence of limb length discrepancies of less than 2 cm was much greater. The male to female ratio was 2:1 (Guichet *et al.*, 1991). Minor degrees of leg length discrepancy are common. Kujala *et al.*, (1987) found an inequality of at least 5 mm in 58 of 141 athletes.

PRINCIPLES OF LIMB LENGTHENING

Osteogenesis is the generation and mineralization of new bone tissue that is stimulated and maintained by the physiological environment associated with fracture repair. Bone is capable of healing by regeneration if favorable conditions exist. In all methods of osteogenic distraction, the bone is simply made to continue to heal as the distraction gap widens.

In the first phase of osteogenesis distraction, an inflammatory response is produced when the bone is surgically cut, by a transection method that preserves the blood supply, called a corticotomy. Following the corticotomy, osteogenic cells travel from the inner layer of periosteum and outer medullary layer to the artificial fracture site. Tensile stresses applied on bone maintain and enhance the process of osseous regeneration.

The regenerating distraction area is initially bridged by fibroblasts which produce longitudinally directed collagen fibrils when distraction is applied. Growth of vascular channels, parallel to the distraction, is also found on the periphery of a central fibrous interzone (Aronson, 1991). Osteoid formation, along the collagen fibrils, forms microcolumns of bone that span the vascularized region from each corticotomy surface to the fibrous interzone. Bone eventually grows across the fibrous interzone when distraction is discontinued and anatomic loading is applied. The osteogenesis produced with distraction resembles longitudinally directed trabeculae seen in epiphyseal plates of endochondral bone (Aronson *et al.*, 1990).

Given sufficiently stable fixation, preservation of the bone vascular supply, and adequate mechanics during distraction and regeneration, the medullary callus fills the gap with new, woven bone, preparing the foundation for the subsequent development of new osteons to connect the fragments.

Physiological factors

Distraction osteogenesis requires preservation of the blood supply to each bony surface by a low energy corticotomy, gradual distraction, and stable fixation which avoids disruption of the microstructures within the osteogenic gap (Aronson, 1994; Frierson *et al.*, 1994). A bone is lengthened by distraction osteogenesis after the cortex of the bone is cut by a corticotomy, with preservation of the peripheral vascular supply from the periosteum. From a practical point of view, it is almost impossible to achieve genuine corticotomy with the preservation of both the intramedullary and the periosteal blood vessels (Yasui *et al.*, 1993).

Many techniques have been devised to assist in the corticotomy including a hand powered intramedullary saw which cuts the cortical bone without damaging the periosteum (Sasso *et al.*, 1993). Any bone lengthening device must be implanted with limited power reaming designed to minimize both heat, which kills bone cells and disruption of the periosteal blood supply.

Ideally, both the periosteum and the intramedullary supply should be preserved during distraction and regeneration. This is possible with external lengthening techniques. However, with a closed technique this may not be possible. Either the intramedullary supply must be disrupted by the placement of an intramedullary nail, or the periosteum by lengthening plates. The issue is then which blood supply should best be preserved.

Regenerate bone formation clearly depends on the periosteum for its blood supply. Ilizarov demonstrated the importance of the periosteum to bone revascularization after fracture (Ilizarov, 1990). He showed that the periosteum participated in the filling of the distraction gap after corticotomy. Kojimoto *et al.*, (1988) have shown, in a study on rabbits, that preservation of the periosteum is necessary if bone lengthening by callus distraction is to succeed. When the periosteum was excised, the regeneration was inadequate to heal the distracted bone. However, scraping of the periosteum had little effect on distraction osteogenesis.

Although it is generally agreed that preservation of the periosteum is necessary, preservation of the intramedullary vascularization may not be essential if the periosteum is sufficiently preserved. Delloye *et al.*, (1990) found no significant difference in the pattern of healing or in the newly formed bone following osteotomy or corticotomy in a group of thirteen dogs. Additionally, bone wax plugging of the intramedullary canal had no major effect on the amount of new bone deposited. However, in the plugged bones only periosteal new bone was evident. This suggests that although preservation of the blood supply from the periosteum is critical, the bone can regenerate with a disrupted medullary blood supply.

Clinical outcome may also depend on the site where the distraction takes place. If the corticotomy is in the mid-shaft, the diaphysis, the bone segments are more stable since the segments are relatively long, closer to equal length, and easier to immobilize with screws and pins. However, researchers have suggested that bone regenerates faster at the metaphyseal region, (the upper or lower third of the bone) because it has a bigger periosteal surface and a richer vascular supply (Price, 1989b). The results of studies comparing metaphyseal and diaphyseal distraction differ significantly. Aronson (1994) suggests that new bone formation and mineralization were better and faster in metaphyseal than in diaphyseal bone. *In vivo* studies on sheep tibia have found no significant difference in torsional strength of elongated bone after metaphyseal and diaphyseal lengthening (Steen *et al.*, 1989). However, movements in the lengthening zone were significantly larger in the metaphyseal than in the diaphyseal region due to the instability of the distraction gap. This motion did not adversely affect bone healing.

According to Wolff's law, which governs all fracture healing, the relative loading pattern of the bone determines the morphologic features of fracture repair. Strain induces osteoblast proliferation. Tissue response is affected by the local environment, particularly oxygen tension. The interfragmentary strain is defined as the ratio of the relative incremental increase of the distraction distance to the gap length before the increase (Chao *et al.*, 1989). According to this theory, a balance between the local interfragmentary strain, the physiological environment and the mechanical characteristics of the callus tissue, determines bone regeneration.

Various tissues can tolerate different maximum tensile strains before failure. Granulation tissue can sustain 100% of strain, fibrous tissue and cartilage can tolerate less, and compact bone can tolerate only 2% of strain (Chao *et al.*, 1989). Strain is inversely proportional to the size of the fracture gap. During the first phase of distraction osteogenesis, when the gap is small, interfragmentary motion that increases the strain more than 100% can prevent granulation tissue from forming. To compensate, small sections of bone near the fracture gap may resorb, enlarging the gap and reducing the strain (Chao *et al.*, 1989).

The effects of bending and shear on osteogenesis are less clear. The effect of pure bending forces is unknown (Paley, 1990). Shear stresses are thought to hinder bone formation, although this has not been proven. Any device designed to assist osteogenic distraction, must not produce more strain across the distraction gap than can be tolerated by the granulation tissue.

A significant factor in promoting periosteal callus formation is the amount of physiological loading at the fracture gap (Aro and Chao, 1993). Simply widening the gap does not necessarily provide a sufficient environment for osteogenesis. The site must also be stimulated by compressive physiological loading. Clinically, some weight bearing will produce axial loading of the regenerating tissue that may stimulate callus formation (Kershaw *et al.*, 1993; Kenwright and Goodship, 1988).

If applied cautiously, many of the principles of fracture healing can be applied to osteogenic regeneration following distraction. However, the ideal level of axial compressive stress during regeneration is not known. In a fracture healing study by Kershaw *et al.*, (1993), a micromovement module was attached to an external fixator that increased axial movement at the fracture site by 50% during walking. This resulted in a significant reduction in time to healing. Experimental work has shown that early micromovement of 1 mm imposed across a 3 mm distraction gap can promote callus formation in sheep.

The effect of dynamization was studied clinically in 80 patients treated with an external unilateral frame. Half the patients used a pneumatic mechanical stimulator that displaced the fragments 1 mm axially at 0.5 Hertz in one session of twenty minutes each day for a maximum period of three weeks. The mean healing time for the group subject to micromovement was 23 weeks versus 29 weeks for those without (Kenwright *et al.*, 1991).

The axial stiffness of external fixation devices tends to be low, varying between 2000 N/cm and 4000 N/cm for most frames and pin configurations. (Chao *et al.*, 1989) This means that partial weight bearing of 200 N causes 0.5 mm to 1.0 mm of axial cyclic movement across the distraction gap. This movement is not confined to the distraction gap. Clinical experience and three dimensional, finite element analysis has shown that motion at the fixation pin to bone interface can lead to bone resorption and loosening of the pins (Chao *et al.*, 1989).

Chao *et al.*, (1989) have demonstrated that four factors most influence the mechanical performance of the pin-bone interface. They are pin geometry and thread design, bone thread preparation, pin insertion technique, and pin-bone stress. The major methods of enhancing external fixation rigidity include increasing the pin diameter, using more pins, decreasing the distance between the external frame and the pin sites, decreasing pin separation, and increasing pin-group separation.

Mechanical factors

Neutral fixation, or latency, for a certain period before distraction is started is sometimes necessary for restoration of circulation in the periosteum and marrow cavity (White and Kenwright, 1991). Latency is the time between the operation and the initiation of distraction. Depending on the local damage done during the corticotomy, latencies from 0 to 7 days preceded successful distraction and production of bone (Aronson, 1991). If the cortex is gently separated with the preservation of the periosteal sleeve, distraction can begin immediately (Aronson, 1991). However, a longer latent period is recommended for older patients, or when lengthening is performed through dense or previously injured bone.

Aronson (1991) suggests that a latent period for either metaphyseal or diaphyseal lengthening sites may not be necessary and too long a latent period resulted in premature consolidation. In the study, eight dogs with no latency had normal union with no premature consolidation. The dogs with 1, 2, and 3 week latent periods had only 62.5 %, 62.5 %, and 12.5% success for bone union without premature consolidation or nonunion. Clinically, however, the importance of a latent period is highly dependent on the individual patient's osseous blood supply and the environment surrounding the distraction site. Any device designed to lengthen bone, must permit the physician and patient to control the latent period.

The rate of distraction should remain in the range of 0.5 mm to 2.0 mm per day averaging 1 mm per day (Ilizarov, 1990). Slower rates allow normal fracture healing to continue and premature bridging or consolidation of the gap occurs. Faster rates seem to outstrip the advancing blood supply, inhibiting mineralization (Aronson, 1991).

Rhythm is defined as the number of incremental distractions over time. Nearly all external lengthening devices currently used are mechanically controlled by the patient turning a screw mechanism. Thus, the distraction rate is broken down into a clinically practical rhythm. Adequate osteogenesis occurs at rhythms ranging from 0.5 mm every 12 hours to 0.25 mm every six hours (Aronson, 1991). Theoretically, the gap should respond optimally if the incremental distraction distance is infinitely small and the rhythm is continuous. Ilizarov (1990) described the advantages of a continuous gradual distraction using a motor driven autodistractor that lengthened at 1 mm per day with a frequency of 60 increments per day. Motorized, computer-controlled external distraction mechanisms are commercially available. Because of minimized disruption of the distraction gap, the bone tends to regenerate faster when auto-distractors are used. As a result, premature consolidation often occurs with this technique unless the rate of lengthening is increased by 50% (Paley, 1990).

Biological tissues exhibit viscoelastic behavior to varying degrees. This time dependent mechanical response of tendons and other soft tissues explains the patterns of stress relaxation after each distraction noticed by patients following an incremental lengthening. Thus, slow continual distraction is better for soft tissues (Leong *et al.*, 1979). More frequent intervals of less distraction may also allow for better soft tissue relaxation due to the viscoelastic behavior of collagenous tissue (Ilizarov, 1990).

The force required to lengthen limbs has been evaluated in recent studies (Brunner *et al.*, 1994; Younger *et al.*, 1994). As shown in Fig. 10.1, the maximum force required to distract bone and soft tissues in 0.25 mm increments varies from 150 N to 700 N. In all cases, the measuring of distraction forces involved instrumenting external fixators with force transducers and calculating the forces required at the distraction zone to allow lengthening.

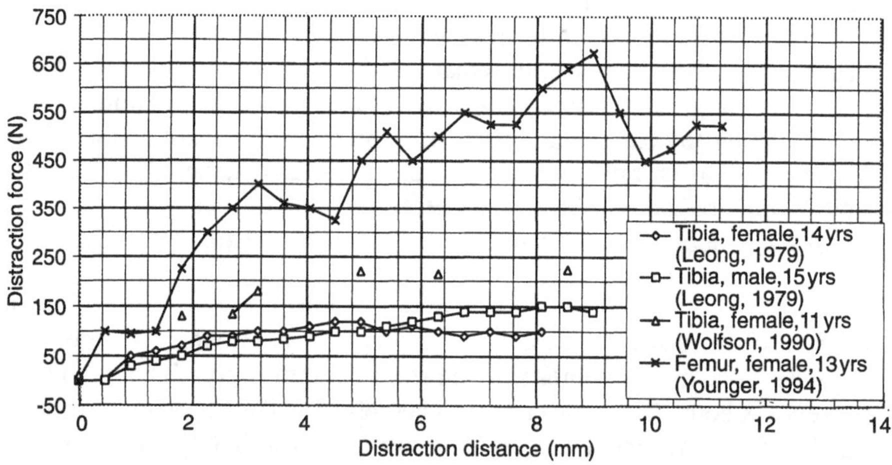

Fig. 10.1 Variability of distraction forces measured in clinical studies.

Verkerke put force transducers on a Wagner external fixator for femoral lengthenings in two patients and found average maximum load of 435 N (Aronson and Harp, 1994). The viscoelastic response of the soft tissue was noted as the most significant reason for the force increase immediately after lengthening. The study suggested that monitoring distraction forces during lengthening may help determine the clinical outcome of procedures (Aronson and Harp, 1994; Younger *et al.*, 1994).

An *in vivo* study on sheep tibiae by Steen *et al.*, (1988) further documented the force required for lengthenings. The decrease in tension of soft tissues was estimated by fitting experimentally obtained data to a five parameter Kevin chain spring-dashpot element model. The model was then used to simulate the relaxation response. The results verified that the force required to maintain a given tissue length decreases with time. The distraction force required to maintain a 1 mm distraction gap will decrease 43% over 24 hours. Due to stretching of collagenous tissue, a residual passive tension of 431 N occurred after 26 mm (12.5%) lengthening in the sheep. Eventually, passive tissue tension is further reduced by active cellular growth of muscle fibers and connective tissue in a process called adaptation (Steen *et al.*, 1989).

In gait analysis studies, dynamic compressive forces as high as four times body weight have been measured by instrumented static intramedullary nails used for fracture fixation. One gait analysis study of instrumented external lengthening devices suggested that forces of this magnitude are not expected during weight bearing by patients undergoing leg lengthening procedures. (Younger, 1994). Surprisingly, in this study, the compressive forces in the frame did not significantly increase during walking. This suggests the presence of an inelastic callus. If the callus were elastic or mechanically insignificant, ground reaction force would transfer to the frame.

Physiological constraints

In earlier reports Kawamura *et al.* (1968) found that the safe limit for lengthening was 15% of the bone's original length. More recently, reports of 50-100% (5-10 cm) are common (Paley, 1991). The rate of regeneration is not necessarily the controlling factor in lengthening rate. The viscoelastic behavior of elongated soft tissues often limits the lengthening rate. Thus if the bone is regenerating faster than the soft tissues can tolerate, premature consolidation is possible. In a recent study by Yasui *et al.*, (1993) the periosteum and the muscle fascia were labeled with metal markers and changes in the position of the markers on 30 rabbit tibias lengthened with dynamic external fixation were monitored. As expected, the elongation of muscle was shown to occur throughout the muscle and not simply at the site of osteotomy.

Summary of factors affecting limb lengthening

The physiological factors effecting the success of lengthening include:

- Preservation of the blood supply--periosteum most critical;
- Metaphyseal vs. diaphyseal distraction--diaphyseal more stable, metaphyseal more vascular;
- Tensile strain during distraction--strain should not exceed 100%;
- Compression during regeneration and healing--dynamization may decrease healing time.

The mechanical factors effecting the success of lengthening include:

- Latency--0 to 7 days recommended (patient dependent);
- Rate--0.5 mm to 2.0 mm per day (1 mm per day most common);
- Rhythm--smaller, more incremental distractions preferred;
- Tensile force during distraction--from 150 N to 700 N (dependent on viscoelastic tissues);
- Dynamic compressive loading--weight bearing is beneficial during regeneration.

CURRENT SOLUTIONS

Equalization by contralateral shortening

Leg lengthening is the treatment of choice for leg length inequality. However, bone shortening is an alternative that is considered for corrections of between 2 and 6 cm in adults of reasonable total height (Kenwright *et al.*, 1991). If a shoe lift has not been successful, closed femoral shortening may be beneficial. It can be performed in mature patients when the discrepancy is known and stable and the desired degree of correction can be obtained precisely. Compared with lengthening, it is a single closed procedure with fewer complications, quicker healing time, and less required rehabilitation (Sasso *et al.*, 1993). Limb shortening is applicable mainly to the femur. Few patients complain about cosmetic appearance after femoral shortening but tibial shortening usually leaves a permanent prominence in the leg. As the limb is shortened, the muscles and soft tissues bulge, making closure of the wound difficult. The short limb is usually thinner, so an increase in the circumference of the larger thigh accentuates the difference. In the tibia, 3.0 cm is the maximum shortening normally possible because of risks of complications including compartment syndrome, neurovascular complications, disfigurement and an unacceptable degree of muscular weakness. Femoral shortening regularly achieves the goal of leg length equality without loss of function. Most of the problems reported in the literature are associated with inadequate fixation and such complications are preventable.

Amputation followed by orthotic management

Once the severity of the discrepancy is measured, the patient and the surgeon determine if lengthening is feasible or if prosthetic management is more applicable. Orthoses are used to equalize moderate leg length discrepancies, to provide joint stability and to compensate for smallness of the foot. A prosthesis is used to equalize leg lengths to improve alignment, to stabilize joints and to improve appearance.

In light of significant developments in limb salvaging procedures, the merit of amputation needs to be discussed. Amputation is advocated for severe cases which would involve numerous operations, entailing long periods in the hospital, and with a high risk of a poor outcome (Fergusson *et al.*, 1987). Although the decision to amputate is always difficult it may be necessary. Many patients adapt better to a prosthesis than had limb salvage been attempted.

The indications for limb salvage procedures are less clear if a major leg-length discrepancy and a gross degenerative deformity exist. Decisions are more difficult in patients in whom congenital joint or soft tissue abnormalities may compromise or preclude attempted leg lengthening. Patients with multiple leg lengthening procedures may achieve length equality, but at the expense of considerable loss of joint and muscle function.

In contrast, an early amputation can result in a predictable disability and correctable function.

Direct surgical lengthening

One stage, direct leg lengthening, is generally regarded as major, demanding surgery with potential risks of serious complications. It should be used with great care in only selected cases. Direct lengthening has the advantage of faster rehabilitation without external fixators. A major and frequent complication of direct lengthening is neurovascular complication (Tjernstrom *et al.,* 1993).

Multiple-stage lengthening is achieved with an internal fixation, non-gradual, lengthening device, the Barnes device (Ensley *et al.,* 1993). The first stage generally lengthens the femur 2.0-2.5 cm, and each subsequent stage obtains 2.0 cm of length. The second stage is not performed until full knee flexion is achieved. The patient wears the brace when out of bed and a cylinder cast at night. Full weight bearing is allowed when clinical and radiological union is evident, usually 6 to 12 months after the operation. Plate removal is not routinely performed. The Barnes apparatus is an entirely implantable stainless-steel internal fixation device consisting of two side plates joined by one longitudinal telescoping rod and one turnbuckle screw. The side plates are forced away from each other when the turnbuckle screw is turned clock-wise. Conversely, the plates can be moved together by turning the screw counter-clockwise. The distraction gap is filled with autologous bone during each surgical lengthening. The Barnes device avoids pin tract difficulties prevalent in external lengthening methods. It is a rigid internal fixation method, thus angular misalignment is rare.

In a recently reported clinical study (Ensley *et al.,* 1993), 20 femoral lengthenings were done with the Barnes device. The average length gained was 5.4 cm and there were no malunions or nonunions. There were two transient perennial palsies, but no permanent nerve injuries. There were no pin tract infections, no knee subluxations and no chronic infections. However, an average of three lengthening operations was required for each patient.

Lengthening with an external device

Although they vary in appearance, all dynamic fixators consist of a small number of components with specific purposes:

- Supporting frame--to provide axial, torsional, and bending support;
- Screws or pins--to anchor the fixator to the bone sections;
- Joints--to connect the screws or pins to the supporting frame;
- Drive mechanism--to provide dynamic distraction of the bone segments.

The two most common types of external fixation supporting frames are uniplanar, sometimes referred to as the cantilever, and circular, most commonly recognized as the Ilizarov device.

Two methods have been used which combined periodic lengthening with bone grafting. The Wagner method was the method of choice through the 1970s. After a transverse osteotomy, the bone is immediately lengthened 2.0 cm to 3.0 cm during the first surgical procedure. Then the bone fragments are distracted by the uniplanar external Wagner fixator at a rate of roughly 1 mm per day until the desired length is achieved. A second operation is done to fill the distraction gap with either autologous or allograft bone graft, and a bone plate applied to hold fragments and the graft together. Finally, after the graft has had time to heal, the plate is removed.

The Wagner device is the first that is small enough and strong enough to allow patient mobility. The Wagner device is a quadrilateral telescoping fixator. The pieces that hold the screws can be tilted through an angle of 40°. Distraction or compression can be applied by turning a knob clockwise or counter clockwise. Three transcutaneous screws are attached to the device at the proximal end and three screws are at the distal end of the lengthening apparatus.

When the Wagner device was first introduced, a bone graft was used to fill the distraction gap. However, after the biology of osteogenic distraction was better understood, following the principles popularized by Ilizarov, the device was used as a gradually lengthening osteogenic distractor and was removed after bone regeneration and healing without bone graft implantation. Many authors have reported on their clinical results using the Wagner device with and without bone grafting. However, only those which distinguished the difference are reported in this chapter, the bone graft method is classified as the 'Wagner method' and the distraction method with the Wagner device is classified as 'Distraction by Wagner device' (Österman and Merikanto, 1991; Coleman and Scott, 1990; Amour and Scott, 1981; Guarniero and Barros, 1990; Tolo and Mears, 1983; Lokietek *et al.,* 1991; Luke *et al.,* 1992; Lai 1991; Nakamura *et al.,* 1991; Dahl and Fischer, 1991; Hood and Riseborough, 1981).

Wasserstein developed a lengthening technique using cortical allografts (Wasserstein, 1990). An osteotomy is performed through the diaphysis of the bone to be lengthened. Then, a dynamic distracting compression ring fixator is applied. Approximately 10 days following surgery the bone is distracted 1 mm to 2 mm daily. A carefully chosen diaphyseal allograft is prepared. The allograft contains no marrow or periosteum and longitudinal cleft is cut into it. It is placed between the bone ends under compression. The longitudinal cleft allows bone regeneration to take place externally and internally. Compression is maintained for 6 to 8 weeks. This method is advantageous in long lengthenings in which prolonged treatment may be intolerable.

Gradual osteogenic distraction has been achieved with both uniplanar (Orthofix) and circular (Ilizarov) devices with external stabilization. The Orthofix device is similar to the Wagner device in that it is also an external, unilateral device with parallel transcutaneous screws connected proximal and distal to the distraction gap. The lengthening mechanism is a screw which is activated by a hexagonal driver. Unlike other external fixation devices which are adaptations of hardware originally designed for rigid, static external fixation, the device was designed to be used as an external, uniplanar distraction device (Guidera *et al.*, 1991; Price and Mann, 1991). The main advantage of the device is that it is stable, easier to attach and less technically demanding for the patient than the Wagner device (Price and Cole, 1990).

The most common circular or ring external fixation device is the Ilizarov device. The Ilizarov external fixator was developed in Russia by Ilizarov in the 1950s (Ilizarov, 1990). The device was developed specifically to allow the surgeon multiplanar adjustments for complicated deformities. The ring fixator provides stable fixation of bone segments to allow functional loading. The typical fixator consists of stainless steel rings that surround the limb and are interconnected by threaded rods. Tensioned Kirschner wires pierce the bone in the plane of each ring and are tightly attached to the rings.

The bone segments are held stable by the tensioned wires within the external scaffolding of rings and threaded rods.

A circular fixator is inherently more stable than a unilateral fixator and therefore more adequately resists deformation forces. Because it encircles the limb, it provides three-dimensional multiplanar support. If deformities occur they can be corrected in any plane (Cattaneo *et al.*, 1991; Harris *et al.*, 1994; Villa, 1991). The inherent stability of the circular fixator combined with its unique ability to correct deformity in all planes has allowed longer lengthenings. Biomechanical analysis of the Ilizarov fixator has shown that its stiffness in axial loading is less than other common fixators. The theoretical advantage of decreased axial stiffness allows cyclic loading of bone during weight bearing which has been shown to enhance fracture healing (Golyakhovsky and Frankel, 1992; García-Cimbrelo *et al.*, 1992; Bonnard *et al.*, 1993; Catagni, 1991).

To lengthen a limb, the bone must be divided. This immediately causes instability which increases as the soft tissues become tight. Since the soft tissue masses are not evenly arranged around the bones they exert uneven forces on the fixation device. Thus, the circular frame devices theoretically provide stability which is not possible with uniplanar devices. However in clinical practice, the complexity of equalizing these forces by balancing the distraction rings and adjusting three to five threaded rods, can be demanding for the patient. Many clinicians and patients prefer the one screw adjustment and the less bulky uniplanar distraction devices.

Intramedullary stabilization has also been used. Early clinical experience with lengthening of the femur over an intramedullary nail was reported by Bost and Larsen. The results of lengthenings of twenty-three procedures on twenty children from 5 to 15 years old were reported. After surgically implanting a small diameter stainless steel intramedullary rod, Steinmann pins were inserted to avoid contact with the intramedullary rod. Immediately following the operation, the patients underwent heavy continuous traction applied to the distal femur at a slow rate until the desired length was obtained. The average length gained was 5.4 cm, the average time required for lengthening was about eleven weeks. Although complications were evident in this early series, mainly because the modern techniques of callus distraction had not been realized, it was apparent that the intramedullary rod had eliminated the usual difficulty of controlling the alignment of the bone fragments (Bost and Larsen, 1956). Additionally, the presence of the rod in the intramedullary canal did not seem to affect the regeneration of new bone.

Since the introduction of stronger, smaller diameter intramedullary nails, than those used by Bost, orthopaedic surgeons are more often combining the advantages of intramedullary stabilization and external dynamic lengthening by gradual distraction. This technique requires additional precautions against infection since the external fixation pin is a direct path for micro-organisms to the intramedullary canal.

COMPLICATIONS OF LENGTHENING METHODS

Despite advances in the surgical correction of limb length inequality, complications continue to plague patients. Limb lengthening has one of the highest complication rates of any orthopaedic procedure. Ilizarov (1990) reported a 5% complication rate with his ring fixator design. DeBastiani reported 14% with the Orthofix uniplanar design (DeBastiani *et al.,* 1987), Wagner reported 45% (Wagner, 1978) and Moseley and Mosca (1988) 142 complications in 63 lengthenings (225%) using the Wagner technique. Enthusiasm for distraction osteogenesis, combined with commercialization of limb-lengthening devices has encouraged surgeons to engage in this field who have not been formally trained in limb-lengthening techniques (Dahl *et al.,* 1994). Dahl plotted the decrease in rate of complications as a function of experience for the three main types of procedures. In this study, the Ilizarov method was used in 84 patients, the DeBastiani method with the uniplanar external fixator in 34 patients and the Wagner technique using autologous, cancellous bone in 22 patients. Complications decrease significantly as experience is gained (Fig. 10.2).

Infection

Pin tract infections continue to complicate lengthenings using external fixation systems. The most frequent complication of limb lengthening by external methods is pin site inflammation. This can lead to soft tissue infection and eventually to osteomyelitis. In a review by Eldridge and Bell, (1991) 22% of 407 Wagner lengthenings had infection, 2% of the 400

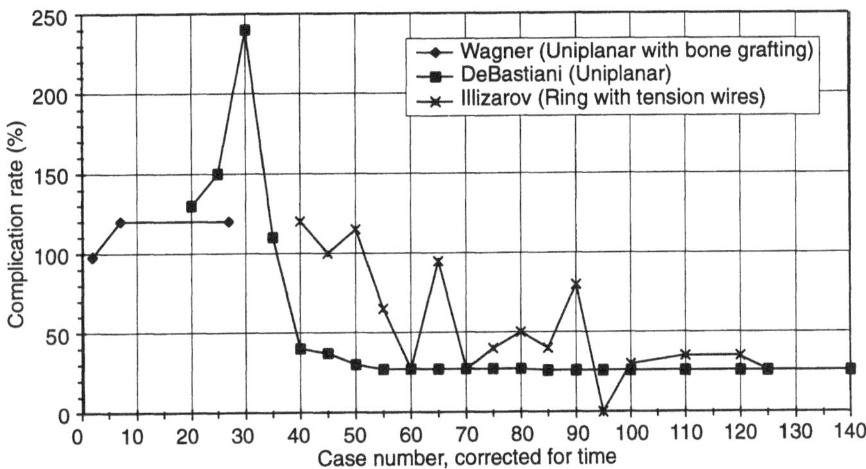

Fig. 10.2 The learning curves of the three procedures (Dahl 1994).

uniplanar, and 10% of the 571 circular fixators had infections. Green reviewed twenty clinical studies from 1942 through 1970 showing that the incidence of complications associated with minor pin tract infections range from 2% to 80%, and major pin tract infections range from 0% to 23% (Green, 1983).

In the USA, infection rate at pin sites has been reported to vary between 8% and 20%. In Kurgan Russia, where the Ilizarov external fixation procedure was pioneered and in U. S. centers with more experience in routine pin care, the incidence of infections is significantly less common (Grant *et al.*, 1992).

Nerve or vascular injury

The neurovascular structures of the extremities are vulnerable to injury both at the time of pin insertion and during distraction. The nerve or vessels may be injured during frame application by transfixion, thermal necrosis, or wrapping around the drill or wire (Pettine *et al.*, 1993). Pin-related nerve injury must be avoided during the operative insertion of the transfixion wires or pins (Young *et al.*, 1994). This type of injury is best prevented by a thorough knowledge of the cross-sectional anatomy and placement of wires in safe anatomical planes (Paley, 1990).

During distraction, the nerve may be stretched too rapidly, causing pain and paresthesia (Eldridge and Bell, 1991). Distraction-related nerve injury occurs much too frequently. It is important to recognize the early signs and symptoms. The patient will usually experience severe pain, then hyperesthesia, followed by decreased muscle strength and finally paralysis. If treated early, paralysis can be avoided (Paley, 1990).

Severe pain

Pain is the most common complaint during limb lengthening. Surgical pain may be quite intense in the first few postoperative days. Contraction of any muscle transfixed by pins is initially painful but resolves within a week or two. The amount of pain increases with the number of osteotomies. During the distraction phase of lengthening, a constant dull aching is often experienced. This is more common in longer lengthenings. The probable cause is stretching of the muscles and nerves. The pain is usually noticed at night and during therapy and walking (Paley, 1990). Associated loss of appetite is also common. This resolves after the distraction period is over. Depression also occurs in some patients (García-Cimbrelo *et al.,* 1992).

Other investigators have characterized patterns experienced by pediatric patients during Ilizarov treatment and found high levels of pain extending over several months, much more than that associated with general orthopaedic pain experience (Young *et al.*, 1993). Limb lengthening by external fixation is a painful process, particularly because of discomfort caused by the localized stretching of tissue by the fixation pins. An implant that obviates external fixation may be less painful simply because the patient does not focus on the transcutaneous pin site.

Pin loosening

Pin tract problems always develop from outside and progress towards the inside. They start with soft tissue inflammation after the pin hole is first drilled. Localized stress during distraction and during loading leads to soft tissue and bone resorption. This causes the pins to loosen and not provide support for the frame. If the resorption is not controlled, infection follows which may eventually lead to osteomyelitis (Paley, 1988).

Device failure

Nearly all external fixation systems are designed with significant mechanical safety factors, since the size of the device is not as constrained as for implantable devices. The material selection is also wider, since external fixation devices are not implanted. Typical external devices are fabricated from non-implantable alloys of aluminum and stainless steel. Given these substantially less constrained conditions, frame failures are rare. However fractures of transfixion wires and screws are commonly reported.

Hip or knee complications

Hip dislocation and subluxation are often reported during and after limb lengthening (Suzuki *et al.*, 1994). Joint stiffness occurs due to persistent contractures or to stiffness of the joint due to the increased pressure on the joint surface during lengthening (Paley, 1990). Functional loading creates an environment in the regenerating bone that enhances ossification and maintains joint motion. As the bone lengthens by distraction osteogenesis, the range of motion in the joints surrounding the regeneration must also be maintained. The vascular, neurologic, and musculo-tendinous soft tissue of a limb must elongate as the bone does. In normal bone growth, the soft tissues are tensioned, stressed, and stimulated to grow during normal movement, maintaining joint motion. If the soft tissues do not lengthen at the same rate as the distracted bone, a relative shortening of the tissues will occur and the joints will become stiff.

Bone fracture after device removal

Patients with a congenital limb length discrepancy are prone to spontaneous fracture after limb lengthening (Luke *et al.,* 1992). Osteoporotic stress fractures in the bone may occur at the normal level. This is due to marked osteoporosis that can develop due to lack of weight bearing, the hypervascular response to distraction, pain, and reflex sympathetic dystrophy (Paley, 1990).

Bone misalignment or malunion

During lengthening there is a tendency for the limb segments gradually to deviate. This is due to an imbalance of muscle forces on different sides of the bone (Paley, 1990). It may be seen as gradual axial deviation of the bone, due to incomplete healing; as a complete healing; as a complete fracture, or as buckling of the bone with some loss in length.

Premature consolidation is a serious complication. This is when the lengthening process fails to progress due to premature healing and mineralization of the regenerated bone (Paley, 1990). This complication increases the risk of fracture, bending or collapse of the callus after the removal of the lengthener. Once the distraction gap begins to consolidate prematurely, osteogenic distraction can not be accomplished without refracture.

A major complication of limb lengthening is delayed formation or maturation of the callus. Delayed consolidation is caused by a variety of factors. The technical factors are traumatic corticotomy, initial diastasis, instability, and too rapid distraction. The clinical factors include infection, malnutrition and metabolic factors. The treatment of delayed unions and nonunions has included bone grafting, fixation with plates or intramedullary nails, fibula osteotomy, pulsed electromagnetic fields, and various other surgical procedures requiring admission to hospital or prolonged immobilization.

Social and psychological complications

A few studies have examined the psychological concerns associated with patients with limb length discrepancies and psychological issues related to surgical lengthening procedures (Eldridge and Bell, 1991). An equalization of limb length, or a change in height may readjust the attitudes of patients towards themselves and others. According to a survey of achondroplastic patients before undergoing lengthening procedures, the most common reasons given for wanting to be lengthened were; to be friends with normal sized people, to be able to do ordinary tasks, to be comfortable, to find a good job, and to drive a car. However, the psychological and social challenges associated with treatment must also be considered. Patients should have a supportive social environment during lengthening (Eldridge and Bell, 1991). The mental stress, prolonged morbidity and incessant pain associated with lengthening and the continual presence of an awkward external lengthening device has led to unusual and detrimental psychological behavior in some reported cases. Complications, set-backs and the fear of eventual amputation compound the psychological reactions. These issues are more evident in children. The requirements of each patient need to be analyzed so that treatment provides the best functional and emotional outcome.

EXPENSE

Limb reconstruction is cost effective when compared with amputation, if lifetime prosthetic costs are also included in the comparison. A recent study compared the total medical expenses of ten patients with tibial nonunions, osteomyelitis, infected nonunions, or bone defects who underwent limb reconstruction with the Ilizarov device and six patients with similar traumatic injuries who underwent amputation. The total cost of the limb reconstructions averaged $59,213. The hospital cost and professional fees for the amputated group averaged $30,148 without prosthetics and $403,199 when the projected prosthetic costs over an average 34 years was included at annual inflation rate of 5% (Williams, 1994).

CLOSED GRADUAL LENGTHENING METHODS UNDER INVESTIGATION

Although a closed method of gradual osteogenic distraction has been sought for many years, a physiologically practical lengthening implant is still needed. A few fully implantable apparatuses have been designed and tested. Hydraulics, mechanical ratchets and electronics are generally used as energy sources. For the most part, these methods are still under development. Although some animal studies have shown the feasibility of some of these devices, no significant clinical use has been reported.

The goals of a totally closed technique of bone lengthening are the following:

- To maintain mechanical alignment and stability of bone throughout the lengthening, during the regeneration of new bone, and during adaptation.
- To provide a lower risk of infection than any current external lengthening method.
- To reduce the pain associated with lengthening by eliminating the transfixion pins.
- To improve the aesthetics of the patient's limb during and after lengthening by eliminating the external fixation frame and scarring associated with pin sites.

Although the prospects of a fully implantable lengthening device are being realized, there are certain added risks associated with closed techniques, including:

- If the device fails, the consequences may be more severe than of an external device, since an implantable device must be surgically removed.
- From the biomaterials perspective, metallosis from metallic wear or other consequences associated with the movement of lengthening mechanisms could potentially be greater than with simple screws or pins.
- The device must be reliably controlled with adequate feedback to assure the lengthening rate and rhythm.

Closed device requiring external actuators--Baumann device

Baumann has reported on the lengthening of 10 dog femurs using an implantable intramedullary extension nail which is driven by a flexible driver (Baumann and Harms, 1977). The driver is passed percutaneously, then down the medullary canal of the femur, and into the center of the device. The percutaneous driver turns a threaded rod which distracts the upper and lower segment of the intramedullary nail. The telescoping device is periodically manually driven throughout the lengthening process. Once the desired length is achieved, the driver is turned counter clockwise and pulled to disengage it from the threaded rod. The flexible driver is completely removed from the body and the incision closed. The distracted bone is then allowed to regenerate.

Extensions as long as 10 cm were achieved in the animals. An elongation of 13 cm was planned on a human, however, the results were not a part of the published study. The flexible driver remains attached to the implant and the driver handle remains outside the body during the lengthening phase which could take many months. Although no infections were reported in the animal study, the flexible driver did provide a means for infectious organisms to travel into the medullary canal. No studies have been published using this device clinically.

Closed femoral ratcheting devices attached to the hip

Bliskonov devised a technique whereby the patient was his own energy source. Movement between the proximal femur and the iliac wing provided the driving force for the lengthening device. Forty-one such lengthening procedures have been reported. Bliskonov describes a variety of complications but states that a desirable result was achieved in all but one patient (Bliskonov, 1984).

Closed electromechanical devices

In the mid-1970s Witt and Jager (1977) published the results of a sheep femur lengthening study using an electronically controlled implantable distracter. The design consist of a power unit, a control unit and a two part distraction plate. This linearly actuated plate attached to the lateral femur and was activated transcutaneously by a radio controlled transmitter. The whole electronic system including a battery pack and the radio controlled linear actuation, was successfully implanted in three sheep. Although the animal study was successful and the device was tested in a cadaver, no follow-up information was published that would indicate that the apparatus was successful in humans.

The original Witt device was too large to fit in the intramedullary canal. However, a more recent electronic linear actuator design by Betz *et al.*, is designed as a lengthening intramedullary nail (Betz *et al.*, 1990). Two versions of this design have been evaluated clinically. Both versions consist of a telescoping, elongating, cylindrical, motorized actuator that is radio controlled outside the body. One version is powered by a battery pack and the other is powered by inductive current. Both require a receiver that is implanted subcutaneously and connected to the motorized actuator by an electronic cable. The device has been successful clinically in four limb lengthenings, three femurs, and one tibia, in three 26 year old patients. The average lengthening was 4.8 cm. This design has tremendous long term potential since the lengthening rate, and rhythm can be precisely controlled electronically. However, before wide-spread clinical use, the efficacy of the electronic components must be proven.

Intramedullary mechanical devices

Guichet *et al.,* (1992) designed a torsionally activated lengthening device which lengthens 0.07 mm each time it is ratcheted at least 20 degrees. The device has been tested mechanically and evaluated *in vivo*. Their *in vivo* results from 10 sheep show that the sheep bone will regenerate if distracted approximately 1 mm per day under torsional ratcheting. Although two devices failed and two additional sheep were lost to follow-up, the remaining femurs regenerated and healed. The implant is currently being clinically evaluated (Paley, 1994).

The Intramedullary Skeletal Distractor (ISKD) is specifically designed to lengthen under physiologically tolerable movement. The ISKD allows small torsional oscillations between the two telescoping sections to be converted to one-way linear distraction (Justin, 1995). As the patient rotationally oscillates the limb, during normal activity, the device gradually distracts. Since the device is designed to lengthen under rotational displacement as small as 3°, nonphysiological rotational movement is not required to achieve distraction. However, rotational oscillations as great as 13° are allowed. An audible click, mechanically activated after each 0.25 mm of distraction, is recognized by a small external electronic monitor worn around the patient's limb. The monitor verifies the lengthening rate and rhythm. After the desired linear distraction, the device locks both linearly and torsionally to function as a static intramedullary nail until bone regeneration is complete. The device has been evaluated in dynamic mechanical tests and *in vivo* in a sheep. The sheep femur lengthened without any assisted manipulation 1.3 mm per day for 27 mm, then the implant functioned as a static intramedullary nail. A clinical study comparing limb lengthening with the ISKD to limb lengthening by external distraction is underway.

SUMMARY

Despite advances which have been made in the surgical correction of limb length inequality, complications continue to plague patients undergoing these procedures. A review of clinical studies using various limb lengthening techniques shows that complications are common for all methods. However, the complication rate significantly decreases with clinical experience. Experience with external lengthening methods has led to significant advances in the understanding of the optimal biologic and mechanical environment required for successful osteogenic distraction. This experience suggests that, in many cases, an attractive alternative to distraction using external devices would be distraction using a monitored, gradually lengthening, intramedullary device. However, no technological advancements in limb lengthening devices can substitute for clinical judgment and experience.

REFERENCES

Alderheri, R. *et al.* (1988) Lengthening of the lower limbs in achondroplastic patients. *Journal of Bone and Joint Surgery*, **70B** (1), 69-73.

Amour, P.C., Scott, J.H.S. (1981) Equalization of leg length. *Journal of Bone and Joint Surgery*, **63B** (4), 587-92.

Aro, H., Chao, E. (1993) Bone-healing patterns affected by loading, fracture stability, fracture tye, and fracture site compression. *Clinical Orthopaedics and Related Research*, **293**, 8-17.

Aronson, J, Harp, J.H. (1994) Mechanical forces as predictors of healing during tibial lengthening by distraction osteogenesis. *Clinical Orthopaedics and Related Research,* **301**, 73-79.

Aronson, J. (1991) The biology of distraction osteogenesis, in *Operative Principles of Ilizarov,* (eds A.B. Maiocchi and J. Aronson) Williams and Wilkins, Milan, Italy. pp. 42-52.

Aronson, J. (1994) Temporal and spatial increases in blood flow during distraction osteogenesis. *Clinical Orthopaedics and Related Research,* **301**, 124-31.

Aronson, J., Good, B., Stewart, C., *et al.* (1990) Preliminary studies of mineralization during distraction osteogenesis. *Clinical Orthopaedics and Related Research*, **250**, 43-49.

Baumann, E. and Harms, J. (1977) The extension nail, a new method for lengthening of the femur and the tibia. *Archiv fur orthopadische und Unfall-Chirurgie,* **90**, 139-46.

Betz, A, Baugart, R. and Schweiberer, L. (1990) First fully implantable intramedullary system for bone lengthening. *Der Chirurg.* **61**, 605-609.

Betz, A.W., Stock, W., Hierner, R., Baumgart, R. (1993) Primary shortening with secondary limb lengthening in severe injuries of the lower leg: a six year experience. *Microsurgery*, **14**, 446-53.

Bliskonov, A. I. (1984) Prodlouzeni stehenni kosti implantouate aparty, *Acta Chirurgia Orthaedicae Czechoslovaca,* **51**, 6, 451-66.

Bonnard, C. *et al.* (1993) Limb lengthening in children using the Ilizarov method. *Clinical Orthopaedics and Related Research*, **293**, 83-88.

Bost, F. and Larsen, L. (1956) Experiences with lengthening of the femur over an intramedullary rod. *Journal of Bone and Joint Surgery,* **38**, 567-84.

Brunner, U. H. *et al.* (1994) Force required for bone segment transport in the teatment of large bone defects using medullary nail fixation, *Clinical Orthopaedics and Related Research,* **301**, 147-55.

Catagni, M. (1991) Lengthening of the tibia, in *Operative Principles of Ilizarov* (eds A.B. Maiocchi and J. Aronson) Williams and Wilkins, Milan, Italy, pp. 288-309.

Cattaneo, R., Villa, A. and Catagni, M. (1991) Lengthening of the femur, in *Operative Principles of Ilizarov* (eds A.B. Maiocchi and J. Aronson) Williams and Wilkins, Milan, Italy, pp. 310-314.

Chao, E.Y.S., Aro, H.T., Lewallen, D.G., Kelly, P.J. (1989) The effect of rigidity on healing in external fixation. *Clinical Orthopaedics and Related Research,* **241**, 24-35.

Codivilla, A. (1905) On the means of lengthening in the lower limbs, the muscles and tissues which are shortened through deformity. *American Journal of Orthopaedic Surgery,* **2**, 353-69.

Coleman, S. and Scott, S. (1990) The present attitude toward the biology and technology of limb lengthening. *Clinical Orthopaedics and Related Research,* **264**, 76-83.

Dahl, M.T. and Fischer, D.A. (1991) Lower extremity lengthening by Wagner's method and by callus distraction. *Orthopaedic Clinics of North America,* **22** (4), 643-49.

Dahl, M.T., Gulli, B. and Berg, T. (1994) Complications of limb lengthening a learning curve. *Clinical Orthopaedics and Related Research*, **301**, 10-18.

DeBastiani, G. *et al.* (1987) Limb lengthening by callus distraction (Callotasis). *Journal of Pediatric Orthopaedics*, **7**, 129.

Delloye, C. *et al.* (1990) Bone regeneration formation in cortical bone during distraction lengthening. *Clinical Orthopaedics and Related Research*, **250**, 34-42.

Eldridge, J.C. and Bell, D.F. (1991) Problems with substantial limb lengthening. *Orthopaedic Clinics of North America* **22** (4), 625-31.

Ensley, N.J., Green, N.E., Barnes, W.P. (1993) Femoral lengthening with the Barnes device. *Journal of Pediatric Orthopaedics*, **13**, 57-62.

Fergusson, C.M., Morrison, J.D., Kenwright, J. (1987) Leg-length inequality in children treated by Syme's amputation. *Journal of Bone and Joint Surgery*, **69B** (3), 433-36.

Frierson, M. *et al.* (1994) 301 Distraction osteogenesis, a comparison of corticotomy techniques. *Clinical Orthopaedics and Related Research*, **301**, 19-24.

Garcia-Cimbrelo *et al.* (1992) Ilizarov technique results and difficulties. *Clinical Orthopaedics and Related Research*, **283**, 116-23.

Golyakhovsky, V. and Frankel, V.H. (1992) Fixator removal and complications of Ilizarov technique, in *Operative Manual of Ilizarov Techniques*, Mosby-Year Book., Inc. St. Louis, Chapter 10.

Grant, A.D., Atar, D. and Lehman, W.B. (1992) Pin care using the Ilizarov apparatus: recommended treatment plan in Kurgan, Russia. *Bulletin Hospital for Joint Diseases*, **52**, 1.

Green S.A. (1983) Complications of external skeletal fixation. *Clinical Orthopaedics and Related Research.* **180**, 109-16.

Grill, F. and Dungl, P. (1991) Lengthening for congenital short femur. *Journal of Bone and Joint Surgery*, **73B** (3), 439-47.

Guarniero, R. and Barros, T.E.P. (1990) Femoral lengthening by Wagner method. *Clinical Orthopaedics and Related Research*, **250**, 154-59.

Guichet, J.M. *et al.* (1992) Mechanical properties of the gradual lengthening nail, *Transactions of the 38th Annual Meeting Orthopaedic Research Society*, **17** (1), 50.

Guichet, J.M. *et al.* (1991) Lower limb-length discrepancy, an epidemiological study. *Clinical Orthopaedics and Related Research*, **272**, 235-41.

Guidera, K., Hess, W., Highouse, K. and Ogden, J. (1991) Extremity lengthening: Results and complications with the Orthofix system. *Journal of Pediatric Orthopaedics,* **11**, 90-94.

Gustilo R.B. and Mendoza, R.M. (1984) Problems in the management of type III (severe) open fractures, *Journal of Trauma,* **24**, 742

Gustilo R.B. and Anderson J.T. (1976) Prevention of infection in the treatment of one thousand and twenty-five open fractures of long bones. Retrospective and prospective analysis. *Journal of Bone and Joint Surgery,* **58A**, 453.

Harris, N.L. *et al.* (1994) Osteogenic sarcoma arising from bony regeneration following Ilizarov femoral lengthening through fibrous dyplasia. *Journal of Pediatric Orthopaedics,* **14** (1), 123-29.

Hood, R.W. and Riseborough, E.J. (1981) Lengthening of the lower extremity by the Wagner method. *Journal of Bone and Joint Surgery,* **63A** (7), 1122-31.

Ilizarov, G. (1990) Clinical application of the tension-stress effect for limb lengthening. *Clinical Orthopaedics and Related Research,* **250**, 8-26.

Justin, D. F. (1995) *In vivo Evaluation and Mechanical Testing of a Physiologically Activated Intramedullary Skeletal Distractor,* Masters Thesis, University of Central Florida, Orlando, Florida.

Karger, C., Guille, J.T. and Bowen, J.R. (1993) Lengthening of congenital lower limb deficiencies. *Clinical Orthopaedics and Related Research,* **291**, 236-45.

Kawamura, B., Hosono, S., Takahashi, T. *et al.* (1968) Bone lengthening by means of subcutaneous osteotomy, experimental and clinical studies. *Journal of Bone and Joint Surgery,* **50A** (5), 851-78.

Kenwright, J. *et al.* (1991) Axial movement and tibial fractures. *Journal of Bone and Joint Surgery,* **73A** (4), 654-59.

Kenwright, J. and Goodship, A.E. (1988) Controlled mechanical stimulation in the treatment of tibial fractures. *Clinical Orthopaedics and Related Research,* **241**, 36-47.

Kershaw, C.J. *et al.* (1993) Tibial external fixation, weight bearing and fracture movement. *Clinical Orthopaedics and Related Research,* **293**, 28-36.

Kojiomto H., Yasui, N., Goto, T. *et al.* (1988) Bone lengthening in rabbits by callus distraction - the role of periosteum and edosteum. *Journal of Bone and Joint Surgery,* **70B** (4), 543-49.

Kujala, U.M. *et al.,* (1987) Lower limb asymmetry and patellofemoral joint incongruance in the etiology of knee exertion injuries in athletes. *International Journal of Sports Medicine* **8**, 214-20.

Lai, K.A. (1991) Treatment of femoral shortness due to nonunion or malunion by distraction device and huckstep instrumentation. *Journal Formosan Med. Assoc.,* **90** (2), 167-71.

Leong, J.C.Y. *et al.* (1979) Viscoslastic behavior of tissue leg lengthening by distraction. *Clinical Orthopaedics and Related Research,* **139**, 102-109.

Lokietek, W., Legaye, J. and Lokietek, J.C. (1991) Contributing factors for osteogenesis in children's limb lengthening. *Journal of Pediatric Orthopaedics,* **11**, 452-58.

Luke, D. L. *et al.* (1992) Fractures after Wagner limb lengthening. *Journal of Pediatric Orthopaedics,* **12**, 20-24.

Mosley C. and Mosca, V. (1988) *Complications of Wagner Leg Lengthening Behavior of the Growth Plate,* Raven Press, New York.

Nakamura, K. *et al.* (1991) Attempted limb lengthening by diaphyseal distraction continuous monitoring of an applied force in immature rabbits. *Clinical Orthopaedics and Related Research,* **267**, 306-11.

Osterman, K. and Merikanto, J. (1991) Diaphyseal bone lengthening in children using Wagner device: long-term results. *Journal of Pediatric Orthopaedics,* **11**, 449-51.

Paley, D. (1994) Personal correspondence.

Paley, D. (1988) Current techniques of limb lengthening. *Journal Of Pediatric Orthopaedics,* **8**, 73-92.

Paley, D. (1990) Problems, obstacles, and complications of limb lengthening by the Ilizarov technique. *Clinical Orthopaedics and Related Research,* **250**, 81-104.

Paley, D. (1991) Biomechanics of the Ilizarov external fixator, in *Operative Principles of Ilizarov,* (eds A.B. Maiocchi and J. Aronson), Williams and Wilkins, Milan, Italy, pp. 33-41.

Patterson, D. (1990) Leg-lengthening procedure: A historical review. *Clinical Orthopaedics and Related Research,* **250**, 27-33

Pettine, K.A. *et al.* (1993) Analysis of the external fixator pin-bone interface, *Clinical Orthopaedics and Related Research,* **293**, 18-27.

Price, C.T. and Mann, J. (1991) Experience with Orthofix device for limb lengthening. *Orthopaedic Clinics of North America,* **22**, 651-61.

Price, C. T. (1989) Limb lengthening for achondroplasia: Early experience. *Journal of Pediatric Orthopaedics,* **9**, 512-15.

Price, C. T. (1989) Metaphyseal and diaphyseal lengthening, in *Instructional Course Lectures,* (ed J.S. Barr, Jr.) American Academy of Orthopaedic Surgeons, Rosemount, IL., pp. 331-36.

Price, C.T. and Cole, J.D. (1990) Limb lengthening by callotasis for children and adolescents, early experience. *Clinical Orthopaedics and Related Research,* **250**, 105-11.

Sasso, R., Urquhart, B. and Cain, T. (1993) Closed femoral shortening. *Journal of Pediatric Orthopaedics,* **13**, 51-56.

Spiro S.A., Oppenheim, W., Boss, W.K., *et al.* (1993) Reconstruction of the lower extremity after grade III distal tibial injuries using combined microsurgical free tissue transfer and bone transport by distraction osteosynthesis. *Ann Plast Surg.,* **30,** 97-104.

Steen, H. *et al.* (1989) Biomechanical factors in the metaphyseal and diaphyseal lengthening osteotomy, *Clinical Orthopaedics and Related Research,* **259**, 282-94.

Steen, H., Fjeld, T. Bjerkreim, I, *et al.* (1988) Limb lengthening by diaphyseal corticotomy, callus distraction, and dynamic axial fixation. An experimental study in the ovine femur. *Journal of Orthopaedic Research,* **6**, 730-35.

Suzuki, S. *et al.* (1994) Dislocation and subluxation during femoral lengthening. *Journal of Pediatric Orthopaedics,* **14**, 343-46.

Tjernstrom, B., Olerud, S. and Karlstom, G. (1993) Direct leg lengthening. *Journal of Orthopaedic Trauma,* **7**(6), 543-51.

Tolo, V. and Mears, D. (1983) Clinical techniques of limb lengthening, in *External Skeletal Fixation,* (ed D.C. Mears), Williams and Wilkins, Baltimore, MD, pp. 601-45.

Villa, A. (1991) Bone lengthening, in *Operative Principles of Ilizarov*, (eds A.B. Maiocchi and J. Aronson), Williams and Wilkins, Milan, Italy, pp. 285-87.

Wagner H. (1978) Operative lengthening of the femur. *Clinical Orthopaedics and Related Research,* **136**, 125.

Wasserstein, I. (1990) Twenty-Five Year's experience with lengthening of short lower extremities using cylindrical allografts. *Clinical Orthopaedics and Related Research,* **250**, 150-53.

Weiland, A.J., Moore, J.R. and Daniel, R.K. (1984) The efficacy of free tisue transfer in the treatment of osteomyelites. *Journal of Bone and Joint Surgery,* **66A**, 181.

White, S.H. and Kenwright, J. (1991) The importance of delay in distraction osteotomies. *Orthopaedic Clinics of North America,* **22**, 4.

Williams, M.O. (1994) Long term cost comparison of major limb salvage using the Ilizarov method versus amputation. *Clinical Orthopaedics and Related Research,* **301**, 156-58.

Witt, A.N. and Jager, M. (1977) Results of animal experiments with an implantable femur distractor for operative leg lengthening. *Archiv fur orthopadische und Unfall-Chirurgie,* **88**, 273-79.

Wolfson, N. *et al.* (1990) Force and stiffness changes during Ilizarov leg lengthening, *Clinical Orthopaedics and Related Research,* **250**, 59-60.

Yasui, N., Kojimoto, H., Sasaki, K., *et al.* (1993) Factors affecting callus distraction in limb lengthening. *Clinical Orthopaedics and Related Research,* **293**, 55-60.

Young, N., Bell, D. and Anthony, A. (1994) Pediatric pain patterns during Ilizarov treatment of limb length discrepancy and angular deformity. *Journal of Pediatric Orthopaedics,* **14**, 352-57.

Young, N., Davis, R.J., Bell, D.F. and Redmond, D.M. (1993) Electrographic and nerve conduction changes after tibial lengthening by the Ilizarov method. *Journal of Pediatric Orthopaedics,* **13**, 473-77.

Younger, A.S.E. *et al.* (1994) Femoral forces during limb lengthening in children, *Clinical Orthopaedics and Related Research,* **301**, 55-63.

11

Success of Surgery on the Anterior Cervical Spine: Smith-Robinson Technique vs Internal Plates

Jamie M. Grooms
John Bianchi
Joan Chan

INTRODUCTION

Spinal cord injuries affect 8,000 to 12,000 Americans each year. To take care of these patients, society spends approximately 2 billion dollars per year (McSwain *et al.*, 1989). Many of these patients have injuries to the cervical spine, often caused by motor vehicle accidents. Over 21% have cervical spine injuries with up to 25% of those dying or becoming paralysed.

There are many different types of surgical procedures advocated for the treatment of cervical spine injuries, the cost of which ranges from $30,000 to $50,000 per patient.

The surgical options for treatment of these types of injuries are diverse. Many procedures use bone grafting, which may use autogenous bone, allograft, or a synthetic material such as hydroxyapatite. All require some form of support or fixation until bone fusion occurs. This fixation can be internal or external. If it is internal, it can be placed anteriorly, posteriorly, or both. External fixation may consist of a halo type brace with screws attached to the skull, a stiff collar, or a soft collar. Many surgeons believe that, even with so many choices of procedure, none has been shown to be particularly outstanding. In this chapter we compare two different approaches to bone fusion of the cervical spine. One of these procedures, developed in the late 1950s, uses bone graft and some form of external bracing while the other, developed in the middle 1980s, uses bone graft supplemented with an internally and anteriorly placed plate and screw system.

An anterior approach to cervical fusion allows for decompression of both the spinal canal and the cervical foramen. Many techniques have been used to alleviate pain, radiculopathy and myelopathy (Zdeblick *et al.,* 1994). The cervical approach has incorporated bone grafts, as described by Robinson and Smith (1955) and Cloward (1958), who first used autograft material in the 1950s. Autograft is considered to be the 'gold standard' for material used for this procedure. Recognizing autograft as the material of choice the surgeon must be concerned with donor site morbidity, its complications and graft quality. Banked bone can provide consistently higher quality tricortical blocks than can be harvested from elderly patients. Tissue banks usually use material from donors between 50 and 55 years of age for weight-bearing applications. Allograft material has been supplied to the surgical community since the late 1970s.

The Robinson technique utilizes a tricortical bone block harvested from the iliac crest. Cloward uses a bicortical dowel taken from the ilium. Bone bank materials have replaced the bicortical dowel with a unicortical dowel harvested from the femoral head, distal femur, and proximal tibia. Due to the increase in procedures and the difficulty in procuring autograft, (ilium bone blocks), surgeons have tried many substitutes. Usually the fibula, but sometimes a rib, is used as a strut graft to achieve arthrodesis or fusion. Stainless steel and titanium plates are used for immediate decompression, spinal stabilization and patient mobilization (Karasick, 1993).

The complications of an anterior versus a posterior approach are similar, but the anterior approach has more severe complications such as graft dislodgement, esophageal problems, kyphotic and degenerative changes above and below the fusion (White and Panjabi, 1990).

Bone substitutes, now entering the market, challenge the gold standard. Material characterization and comparison of properties with those of natural bone must be considered before any bone substitute is used. Porous hydroxyapatite has very good osteoconductive properties but its greater Young's modulus of elasticity can lead to stress shielding and bone resorption (see Chapter 1). This material is not suitable used alone, for cervical procedures. In animal models for anterior cervical fusions hydroxyapatite had a high failure rate, in 10% a fibrous gap, in 29% collapse and in 14% extrusion (Zdeblick *et al.,* 1994).

In summary, cervical fusion procedures continue to be modified in both technique and materials and such changes must be carefully evaluated in animal models before clinical use.

BACKGROUND

Indications

There are numerous reasons for surgical intervention to treat injuries or disorders of the spine or spinal segments. They include trauma, degenerative disc, herniation, failed fusion after discectomy, tumor, osteomyelitis, spondylosis, Sutterlin has organized the long list of conditions requiring surgical intervention into five surgical goals (Sutterlin, 1995).

1. Deformity correction;
2. Decompression of neural elements;
3. Debridement or excision of pathology;
4. Stabilization of the spinal column;
5. Pain control.

These goals can help to define the surgical method of choice. A patient with a tumor in a vertebral body may be in excruciating pain and experiencing some form of neurological deficit. The surgeon following this patient would have three immediate surgical goals: decompression of neural elements, debridement or excision of pathology, and pain control.

Sometimes the consequences of the surgical technique create another surgical goal. In such an example, a tumor must be removed, which may involve one or more corpectomies. Once these bodies are removed, the spine will be unstable out of the prone position; it must be stabilized. This may require a bone graft or some other type of vertebral body spacer. If the region is large enough and still unstable with simply a vertebral body spacer, supplemental rigid fixation is a necessity. This fixation may consist of a soft or rigid collar, an external metal halo brace, or internal fixation such as an anterior or posterior plate.

Complications

Anterior cervical spinal fusions have been reported to benefit the patient in alleviating pain. Improvement was found in over 91% of the patients. However, this high percentage improvement is accompanied by 29% complications. The complications associated with the anterior cervical fusion procedure include those from bone-grafting, from surgical dissection, from neurological injury, and those arising from postoperative management (Tew and Mayfield, 1976; Whitecloud and Dunsker, 1993).

A major goal of the anterior cervical procedure is to obtain solid fusion. Fusion or arthrodesis occurs when there is visible bone bridging with no movement during flexion and extension of the neck when viewed on the radiograph (Smith and Robinson, 1958; White *et al.,* 1973). Depending on the evaluator, the time period for fusion has been reported as from less than 3 months (Zdeblick and Ducker, 1991) to 20 weeks (Brown, *et al.,* 1976). If fusion takes place within 52 weeks, it is called 'delayed union'. If fusion is not achieved at 52 weeks, non-union is reported (Zdeblick and Ducker, 1991; Brown *et al.,* 1976).

The most common graft problems are those of graft extrusion and vertebral collapse. Rejections occur very rarely (Tew and Mayfield, 1976). Graft extrusion, or dislodgement, is defined when the bone graft is displaced, more than 2 mm (Zdeblick and Ducker, 1991; Brown *et al.,* 1976). Extrusion usually leads to delayed union (Tew and Mayfield, 1976). If the graft is displaced forward, the esophagus and trachea can be compressed. If displaced backward, the spinal cord may be compressed (Whitecloud and Dunsker, 1993). Only if cord compression is present is reoperation necessary (Tew and Mayfield, 1976). Collapse or interspace narrowing occurs when the graft decreases in height by more than 2 mm or 30% (Zdeblick and Ducker, 1991; Brown *et al.,* 1976). If there is a 5° angle change associated with the change, it can lead to kyphosis (White *et al.,* 1973; Zdeblick and Ducker 1991; Brown *et al.,* 1976).

Neurological complication is the type which results in damage to the neuroelements (Whitecloud and Dunsker, 1993). Complications are rare, but can be serious (Tew and Mayfield, 1976). They include quadraplegia, incontinence, and/or numbness. Injuries may result during osteophyte removal or during graft insertion.

During surgery, there are always risks from dissection. In anterior cervical fusion, the spinal cord, nerves, and esophagus may be damaged by blades, drills, or retractors. The most common sequalae are hoarseness, dysphagia, and vocal cord paralysis (Whitecloud and Dunsker, 1993). In addition, infection and hematoma may arise in the wound in the opened area (Tew and Mayfield, 1976; Whitecloud and Dunsker, 1993).

Improper and incorrect diagnosis may lead to unnecessary operations. Fusing the wrong level has also been reported (Whitecloud and Dunsker, 1993).

Complications of the anterior plating systems

One of the problems in using plating systems, especially screws, is that they must be placed in an area where there are nerve roots, arteries, and the spinal cord. These screws must be placed without hitting any of these anatomical hazards. This is especially true when using bicortical screws that thread into the posterior cortex of the vertebral body and may protrude into the spinal canal.

When using a plate, it is important to recognize that bony fusion must occur before the plates or screws fail by fatigue.

When the systems do fail it is usually due to one, or a combination, of the following:

- *Screw backout.* This failure is most common in systems in which the screw does not lock to the plate. It is especially evident if the screw is not placed parallel to the vertebral body end plate, but placed into the disc space. If the screw backs out far enough, revision surgery will be needed to remove it.

- *Screw/plate backout.* If the screw is locked to the plate, the entire screw plate assembly can back out. This type of failure is not common, but if it occurs, it requires revision surgery to repair it.

- *Screw breakage.* The screws are under high stresses and it is not uncommon for them to break. The stresses are especially high if a screw is not placed parallel to the vertebral body end plate but placed into the disc space. Revision surgery is not usually necessary, especially if the fusion has advanced far enough to stabilize the area and the anatomy around the broken screw is not compromised.

- *Plate breakage.* If the plate is thin and the stresses high, the plate can fail. This is especially true of the thin plate, shown in a radiograph by Ripa *et al.* (1991). This is an uncommon complication but may require revision surgery.

Allografts vs. autografts

The horseshoe shaped tricortical bone implant taken from the iliac crest, as proposed by George Smith and Robert Robinson, has been reported to yield improved results in the cervical spine (Smith and Robinson, 1958). However, complications at the iliac donor site have been known to occur (Whitecloud and Dunsker, 1993). These complications include hematoma, infection, nerve damage, and drainage; all result in donor site pain for the patient (Rish *et al.,* 1976; Zdeblick and Ducker, 1991; Wittenberg *et al.,* 1990; Young and Rosenwasser, 1993; Whitecloud and Dunsker, 1993). To eliminate these problems, implants of allograft iliac bone have been used.

Studies have shown that for surgery of only one vertebra, fusion using allograft was similar to that with autograft. In 1976, Brown *et al.,* found autografts had 97% fusion and allografts 94% at each individual level grafted (Brown, *et al.,* 1976). They concluded 'no significant difference' between the two percentages of union. In 1991, Zdeblick showed that for single-level surgeries, the two grafts had the same results, 95% (Zdeblick and Ducker, 1991). Brown found fusion for allografts and autografts in multilevel fusions to be similar, but Zdeblick reported otherwise, 45% for allografts. In addition, partial union and delayed union percentages were not much different for Brown, but Zdeblick reported more delayed unions with allografts. Despite this, Zdeblick found that there was no difference in pain relief from injuries, which is of most concern to patients. Rish *et al.,* (1976) and Lunsford *et al.,* (1980) also found equivalent results both clinically and radiographically for both autograft and allograft bone.

Collapse and extrusion occured more frequently with allografts (Zdeblick and Ducker, 1991; Brown *et al.,* 1976). However, for single level procedures, collapse and extrusion results were similar, 16% and 17% (Brown *et al.,* 1976). The average age of these patients was 45. It is probable that for older patients, autografts would have a higher collapse rate than allografts.

For mechanical stability, good compressive strengths are necessary (Zdeblick and Ducker, 1991; Wittenberg *et al.*, 1990). Young and Rosenwasser (1993) tested fibular allografts, their data showed 92% fusion compared with 88% for iliac autografts, both with satisfactory clinical results (Young and Rosenwasser, 1993). The fibulae, according to Wittenberg, had a mean compressive strength of 5070 N versus iliac bone's strength of 1150 N, and rib's of 452 N, with all of these grafts providing adequate support for spinal loads (Wittenberg *et al.*, 1990). The fibula is strong because it consists mainly of cortical bone. The iliac crest derives strength from its cortical content; the ability for bone growth due to its cancellous region (Cotler and Cotler, 1990). Sedlin and Hirsch (1966) found there was little effect on mechanical properties after freezing an allograft. They reported that samples frozen at -20° C for 3 to 4 weeks had an increase in strength of 5%. Sterilization with ethylene oxide did not affect the strength (Wittenberg *et al.*,1990). The main objection to allograft use is the chance for rejection and infection. However, iliac allografts have been successfully used for anterior cervical spinal fusion. For single-level operations, allografts are adequate with good clinical results (Zdeblick and Ducker, 1991; Brown, *et al.*, 1976; Rish, *et al.*, 1976). Besides eliminating donor site pain, there are also advantages of decreasing operation and hospital time. The size and shape of the graft can also be more versatile and it is more readily available (Zdeblick and Ducker, 1991; Brown, *et al.*, 1976; Young and Rosenwasser, 1993).

Types of allografts which can be used, other than those from the ilium, include those from the ribs and fibulae. The use of iliac crest allograft is emphasized here because it has been more widely used and has a better mixture of cortical and cancellous bone than does iliac autograft.

Biology of bone (tricortical iliac crest block)

Osteoconductive implant materials may be natural or synthetic, cancellous bone and porous hydroxyapatite are such examples. These materials provide scaffolding for the production of new bone. Progression of the healing process is dependent on the host's ability to form new bone and may be impeded by age, smoking and disease. Osteoinductive materials can provide increased ability for local bone formation. Apart from autogenous bone, demineralized bone is the only inherantly osteoinductive material available.

The inductivity of demineralized bone is a function of particle size and exposure of bone morphogenic proteins (BMP). The optimal particle size range is 250 - 850 microns. The exposure of BMP is accomplished by 0.5-0.6N HCl acid treatment which removes the bone mineral, and, by thus exposing BMP results in osteoinduction. This may explain why non-demineralized cortical bone implants rarely go through complete resorption. As the host slowly remodels the cortical implant the BMPs are exposed as mineral is removed. This keeps the bone healing process active. Combining cancellous and cortical bone permits bone healing to proceed along two different pathways, i.e. osteoconduction and osteoinduction.

The iliac crest block in itself is a composite material. Cancellous bone is bonded to cortical bone on three sides. Cancellous and cortical bone are similar in remodeling at the beginning but differ greatly in the middle and end phases. The initial response of hemorrhage, inflammation and vascularization is the same for both types of bone (Karasick, 1993). From this point on, cancellous and cortical bone take different pathways for remodeling.

Cancellous bone is an osteoconductive scaffold that allows for rapid vascularization. With rapid vascularization osteoblasts lay down host bone throughout the cancellous scaffolding. The resorptive phase follows, osteoclasts removing the bony scaffold, leaving host bone. Cancellous implants can be completely replaced with host bone, in 3 to 12 months, depending on the size of the implant.

Cortical bone's geometry does not allow similar rapid vascularization, it proceeds through a resorptive phase first. Healing by this process takes more time to achieve union and may never completely be replaced by host bone. To achieve union cortical bone may take up to one year. Union is achieved as osteoclasts lead resorption across the interface between implant and host. The trailing edge of the resorption area is associated with osteoblastic activity, producing viable host bone.

Biomechanics of bone (tricortical iliac crest block)

The biological characteristics of cancellous and cortical bone seem to have a direct affect on the biomechanical incorporation dynamics of the tricortical implant. This has not been specifically reported in the literature studied, but extrapolation of reported data on the incorporation rates of cancellous and cortical bone does suggest that these effects are real. The cancellous portion of the tricortical implant should begin to increase in strength as the cortical portion decreases. The cortical portion's resorptive phase should stabilize between 12 and 18 months and then it should begin to increase in strength. Thus the use of a material with biomechanical characteristics that are between that of pure cancellous and cortical bone should increase the long term probability of successful fusion.

Biology of instrumentation (stainless steel and titanium)

Stainless steel and titanium are inert materials which are morphologically fixed in place. Materials need to be fixed so that stress shielding is minimized, preventing bone resorption and implant loosening. These inert materials will ultimately be encapsulated by fibrous tissue, the thickness of which will determine long term success of the implant. If the fibrous capsule continues to thicken it will lead to loosening and failure. Interfacial movement and associated tissue response is responsible for the thickening of the capsule. In most cases this process is limited and if the implant is left within a thin capsule it will remain stable.

Implant biomechanics

The objective of attaching rigid fixation to the spine is to enhance the mechanical stability of the spine and minimize motion in the fusion zone. Plating systems minimize motion by careful attention to the stiffness of the system. Stiffness must be maximized such that strain in the unstable segment is minimal. However, stiffness must approximate that of the host bone to prevent the stress shielding which causes bone to resorb and the plate to fail (Chapter 1).

Fixation plates also enhance the mechanical stability of the spine. A load directed at an angle from the longitudinal axis of a cylinder causes increasing instability as the length of the cylinder increases. As the length of the bone graft in the cervical spine increases, the greater is the need for supplemental support to keep it from extruding. Graft extrusion is common in unplated procedures but is almost eliminated by the attachment of a plate (Sutterlin *et al.*, 1988).

Although the anterior plating procedure is more common and is preferred over posterior plating, there are biomechanical advantages to placing plates posteriorly (Sutterlin *et al.*, 1988; Coe *et al.*, 1989). The anterior procedure may be more popular for anatomical reasons. It is important to recognize that posterior plates offer some benefits that may outweigh this perceived disadvantage. It has been shown that the strain recorded posteriorly and even anteriorly is smaller in cadaveric spines with implants placed posteriorly than in the same model with plates placed anteriorly. These strain readings were recorded when the spine was subjected to flexion/extension loading (Sutterlin *et al.*, 1988; Coe *et al.*, 1989).

REVIEW

This chapter is a collection of data from articles that dealt with two specific clinical cervical spine procedures. The first was anterior cervical spine fusions that were performed using the Smith-Robinson technique. The other concerned cervical spine fusion that used internally placed plate and screw systems. Nineteen references provided data on the Smith-Robinson technique, and seven on the internal plating technique. These are listed in Table 11.1. The surgical procedures are described in detail in Smith and Robinson (1958); Wood and Hanley (1992) and Zdeblick (1993).

The Smith-Robinson technique

The most common reason for performing an anterior cervical spine fusion operation is to relieve chronic (greater than six months) neck, arm and shoulder pain, which cannot be treated by conservative means such as physical therapy or medication and which has been concluded to be a result of trauma, disc herniation, or spondylosis.

Three main approaches exist for anterior cervical spine fusions. The main difference is the shape of the graft to be inserted into the disc space to achieve fusion. The Smith-Robinson uses a horseshoe shaped graft (Robinson *et al.*, 1962), Cloward, a dowel (Cloward, 1958) and Bailey-Badgley, a strut (Bailey and Badgley, 1960). Each shape was tested in a study by White and Hirsch (1972) for compressive load. The Smith-Robinson had the highest load to failure, 344 N. The Cloward had a 188 N

Table 11.1 Reference citations

Smith-Robinson technique	
Brodke, D.S. and Zdeblick, T.A.	1992
Brown, M.D. *et al.*,	1976
Clements, D.H. and O'Leary, P.F.	1990
Connolly, E.S. *et al.*,	1965
DePalma, A.F. *et al.*,	1972
Emery, S.E. *et al.*,	1994
Goffin, J. *et al.*,	1989
Gore, D.R. and Sepic, S.B.	1986
Riley, L.H. *et al.*,	1969
Rish, B.L. *et al.*,	1976
Robinson, R.A. *et al.*,	1962
Smith, G.W. and Robinson, R.A.	1958
White, A.A., III and Hirsh, C.	1972
White, A.A., III *et al.*,	1973a
White, A.A., III *et al.*,	1973b
Williams, J.L. *et al.*,	1968
Wittenberg, R.H. *et al.*,	1990
Young, W.F. and Rosenwasser, R.H.	1993
Zdeblick, T.A. and Ducker, T.B.	1991
Internal plating technique	
Aebi, M. *et al.*,	1991
Goffin, J. *et al.*,	1989
Randle, M.J. *et al.*,	1991
Ripa, D.R. *et al.*,	1991
Seifert, V. and Stolke, D.	1991
Suh, P.B. *et al.*,	1990
Tippets, R.H. and Apfelbaum, R.I.	1988

compressive load to failure and the Bailey-Badgley type 195 N. For the combined graft and vertebral body, White *et al.*, (1973) found loads of 309 N, 257 N, and 236 N for the three systems respectively. They considered all to be acceptable.

Lunsford *et al.*, (1980) found no significant difference in results between the Smith-Robinson and the Cloward methods. Since the Smith-Robinson horseshoe technique is able to support the largest loads, it will be studied here in further detail. The Cloward dowel is shaped so that the cancellous bone is between two ends of cortical bone. If it is subjected to too high a load, the dowel may fracture in the less dense cancellous region. In addition, the Cloward technique has a lower fusion rate (Tew and Mayfield, 1976).

In addition to the bone graft, some type of eternal fixation is used. This may simply consist of a soft, stiff collar, or a halo brace that has screws which attach to the skull.

The modified Smith-Robinson technique

The Smith-Robinson procedure has been modified by reversing the placement of the graft. Instead of inserting the cancellous side first, the cortical end is placed posteriorly. The higher strength of the cortical portion near the spinal canal is believed to increase fusion rates and decrease graft extrusion and collapse (Whitecloud and Dunsker, 1993).

Anterior plating system

There are three different plating systems, all are in multiple sizes.

1. The Caspar plate (Aesculap® San Francisco, USA) is made from stainless steel or titanium. The screws are designed for unicortical or bicortical purchase. There is not a screw to plate locking feature.

2. The Orozco plate (Aesculap® San Francisco, USA) is a thin stainless steel plate. The screws are designed for a unicortical or bicortical purchase. There is not a screw locking feature. These plates may be cut to fit.

3. The Morscher plate (Synthes® Pennsylvania, USA) is of commercially pure titanium. The screws are designed for unicortical purchase only. This system is the only one with a design which locks the screw to the plate.

Evaluation

Evaluation includes an interview as well as a physical examination and radiographs, which will show changes in the vertebrae, such as graft extrusion, interspace narrowing, and fusion. The sucess of each operation is based on follow-up evaluation, which should combine the opinions of both the patient and the evaluator. The two most common rating systems are those defined by Odom, *et al.*, (1958) and Robinson *et al.*, (1962).

Odom's

Excellent:	No complaints; patient is able to go back to work
Good:	Occasional complaints; able to work
Satisfactory:	Improvement, but still limited in activities
Poor:	Same or worse

Robinson's

Excellent:	All preoperative symptoms relieved; abnormalities same or improved
Good:	Minimal preoperative symptoms; not limited in activities
Fair:	Definite relief but still abnormalities exist
Poor:	Symptoms and signs unchanged

There are also many other systems used such as that in Gore and Sepic (1986).

Excellent:	90% pain relief
Good:	75% relief
Fair:	50% improvement
Poor:	Less than 50% improvement

RESULTS

Table 11. 2 shows the details of the patients studied in the references analyzed (Table 11.1). Improvement is shown in Fig. 11.1 and fusion in Fig. 11.2. Differences in single and multilevel procedures are shown in Figs. 11.3 and 11.4. In Fig. 11.5 the full range of complications, which may be related to the donor site, the surgical procedure or the hardware are shown. In Figs. 11.6-11.9 the data are represented chronologically allowing the effect of surgical and materials optimization to be seen.

Table 11.2. Patient profile

Patient Profile	*No Plate*	*Plate*
Avg. Age	45	41
Age range	(17-75)	(13-84)
Male	0.47	0.74
Female	0.53	0.26
% single-level	0.57	0.44
% multi-level	0.43	0.56
Avg. follow-up (mos.)	31	26

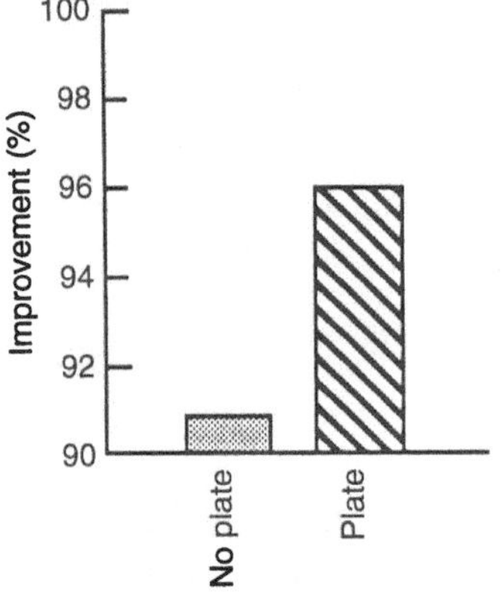

Fig. 11.1 Percentage improvement of plate vs no plate method.

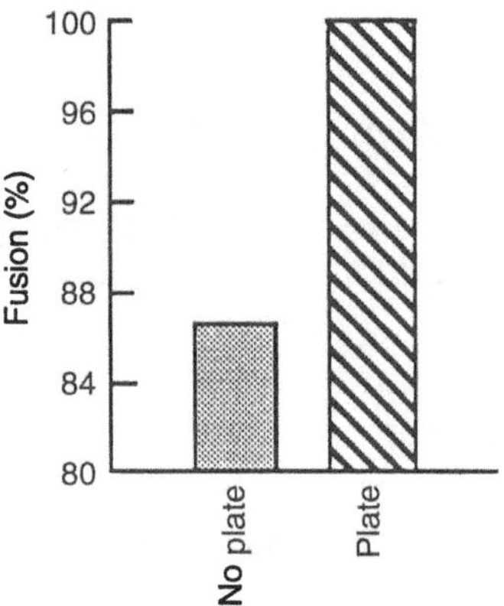

Fig. 11.2 Percentage fusion of plate vs no plate method.

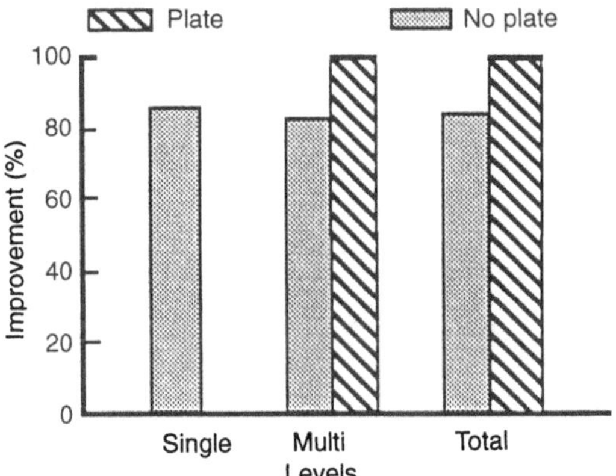

Fig. 11.3 Percentage improvement of plate vs no plate system.

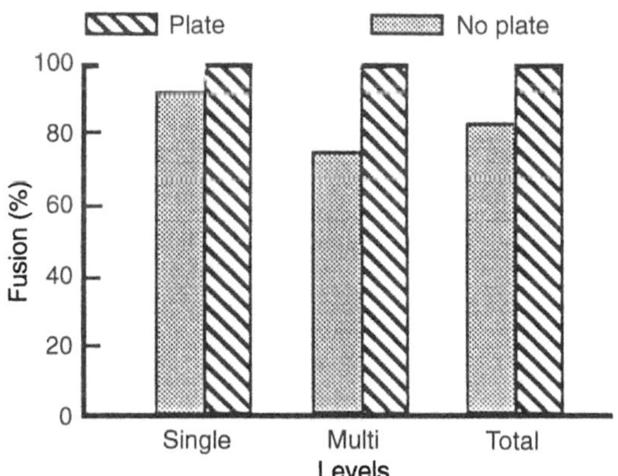

Fig. 11.4 Percentage fusion of plate vs no plate systems.

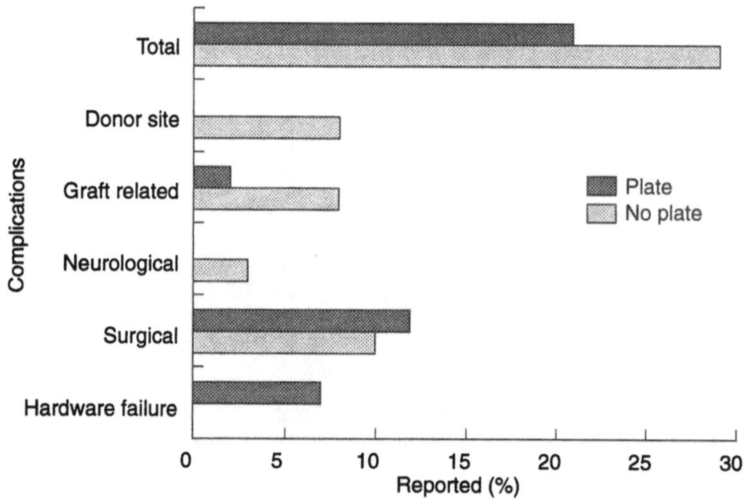

Fig. 11.5 Complications for the plate and no plate system.

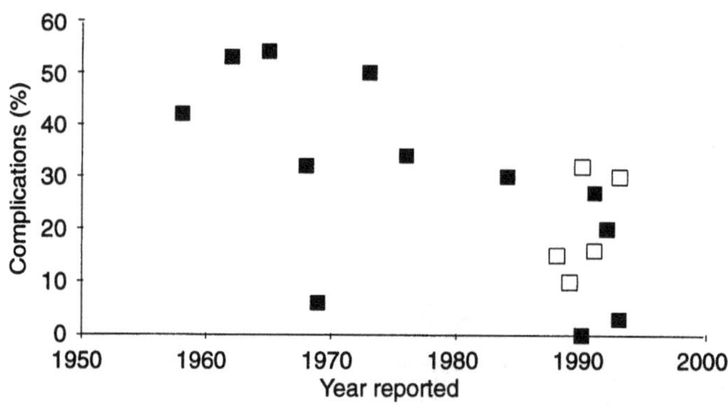

Fig. 11.6 Percentage complications versus the year of study for a plate (□) and no plate (■) system.

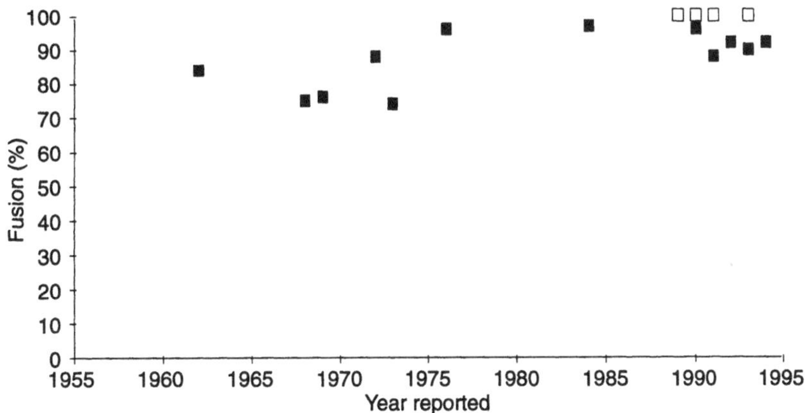

Fig. 11.7 Percent fusion versus the year reported for a plate (□) and no plate (■) system.

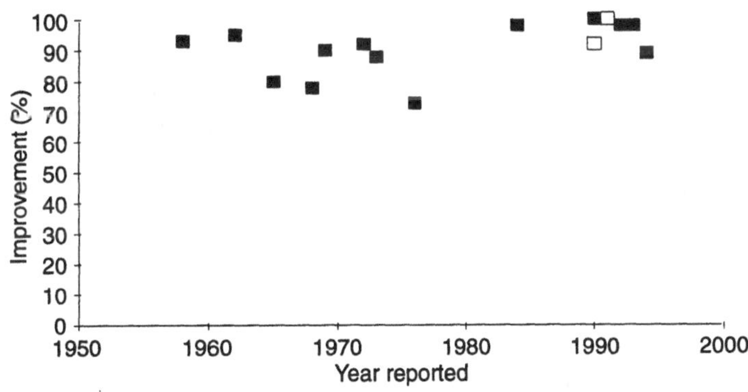

Fig. 11.8 Percent improvement vs. year reported for a plate (□) and no plate (■) system.

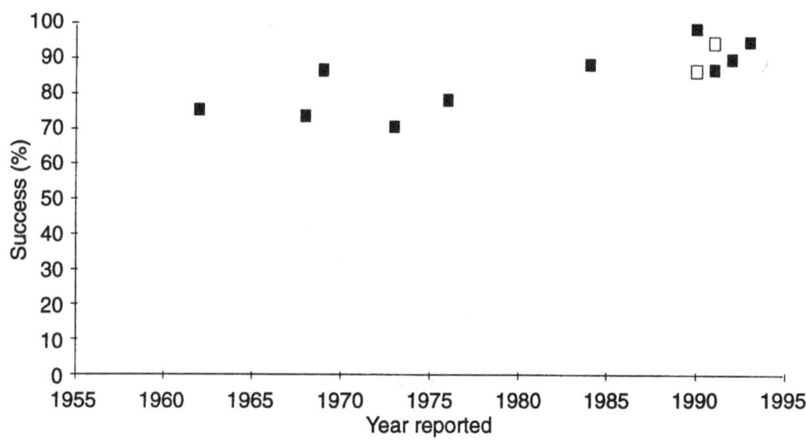

Fig. 11.9 Percent success vs. the year reported for a plate (□) and no plate (■) system.

The average age of the patients was in the mid-40's, with an age range of 13-84. There was an even distribution of single level fusions and multi-level fusions (see Table 11.2). The most common level of operation for single level fusions is the C5-6, for multi-level surgery, C5-6 and C6-7. The sex of the patient was similar for anterior cervical fusions without a plate, but three times more males underwent surgery with a plate. No significant finding relating to the sex of the patient was found. There was not enough information to relate the indications of surgery to the outcome.

A major problem encountered in organizing data from the literature was the use of different evaluation methods. In order to compare data, we used an improvement/no improvement system, where improvement was defined as resulting in benefit from surgery. Another problem was that the articles concerning a plating system focused their study on fusion, whereas the "no plate" articles emphasized pain relief. According to the plating results, 71% of the patients were reported to obtain fusion at 3 months and 100% by 6 months. There was not enough information concerning fusion times for the surgeries without a plate, therefore, only the percentages of patients with fusion were compared.

The average improvement and fusion by use of the Smith-Robinson technique for anterior cervical fusions without a plate were 91% and 87% respectively. The values with plating system were reported to be 96% improvement and 100% fusion (Figs. 11.1 and 11.2). There is disagreement as to whether fusion and improvement correlate. Earlier studies by Robinson *et al.*, (1962) and DePalma *et al.*, (1972) said there was no correlation, whereas in 1993, Newman (1993) changed 69% of unimproved non-unions to unions resulting in improved conditions. It has been suggested that in successful non-unions, there is probably enough stability in the bone graft to give pain relief. However, for long term stability, fusion is desirable. If there is union, the probability of complications resulting from collapse of the graft is lower than without (White *et al.*, 1973). If collapse does not occur after fusion, relief is greater than if collapse occurs (Zhang *et al.*, 1994). However, Riley *et al.*, (1969) found no relationship.

Without a plate, fusion occurred more frequently in single level operations (92%) than multilevel ones (75%), yet relief (85%) was similar. A higher pseudarthosis rate is expected with multiple levels. Single and multilevel fusions with a plate were both reported to be 100% successful (Figs. 11.3 and 11.4). Separate improvement results of a plate in single and multilevel fusions were rarely given, presumably considered to be also 100%.

For a plate versus no plate, the complication rates were similar, 29% and 21% respectively (Fig. 11.5). With time the complication percentages decreased dramatically (Fig. 11.6). This may be attributed to better reporting or to the refinement in operative technique with experience. The percentage fusion increased slightly with the percentage of patients benefitting from surgery remaining fairly constant (Figs. 11.7 and 11.8). The overall succes rate is shown in Fig. 11.9.

The modified Smith-Robinson technique has only been used within the last few years, the earliest results we obtained were from 1992. From these few studies, the modified technique does show high fusion (91%) and improvement rates (95%) and low complication rates (12%). However, long term results are needed before conclusions may be drawn.

CONCLUSIONS

We conclude that the success of an anterior cervical fusion depends equally on fusion, on physical improvement and on complications. Therefore, we have defined the percentage of success by the following formula:

success = (% fusion)/3 + (% improvement)/3 + (100-% complications)/3

The percentage of success did improve with time (Fig. 11.9). The success of the Smith-Robinson procedure without a plate was found to be 84%. The average success of the standard technique using data from 1958 to 1991 was 82%, and for the modified in 1992 and 1993, 93%. Using only the standard method results from 1984 to 1991, the success was calculated as 91%. Success for a plate was 91%.

Although the percentage success for procedures with and without an internal plate were comparable, we conclude that to achieve long term stability, fusion is necessary. Since plating systems are seen to have high fusion rates, plates should be used for multilevel surgeries whereas the standard Smith-Robinson technique without a plate is sufficient for single level procedures.

REFERENCES

Aebi M., Zuber K., Marchesi D. (1991) Treatment of cervical spine injuries with anterior plating; indications techniques, and plating. *Spine* **16**, 38-45

Bailey, R.W. and Badgley, L.E. (1960) Stabilization of the cervical spine by anterior fusion. *J. Bone and Joint Surg.* **42A**, 565-94.

Brodke, D.S. and Zdeblick, T.A. (1992) Modified Smith-Robinson procedure for anterior cervical discectomy and fusion. *Spine* **17** (10S), 427- 30.

Brown, M.D., Malinin, T.I. and Davis, P.B. (1976) A roentgenographic evaluation of frozen allografts versus autografts in anterior cervical spine fusions. *Clin. Orthop.* **119**, 231-36.

Clements, D.H., and O'Leary, P.F. (1990) Anterior cervical discectomy and fusion. *Spine* **15**(10), 1023-25.

Cloward, R. B. (1958) The anterior approach for removal of ruptured cervical discs. *J. Bone and Joint Surg.* **15**, 602-17.

Coe, J.D., Warden, K.E., Sutterlin, C.E., McAfee, P.C. (1989) Biomechanical evaluation of cervical spinal stabilization methods in a human cadaveric model. *Spine* **14**(10), 1122-31.

Connolly, E.S., Seymour, R.J. and Adams, J.E. (1965) Clinical evaluation of anterior cervical fusion for degenerative cervical disc disease. *J. Neurosurg.* **23**, 431-37.

Cotler, J.M., and Cotler H.B. (1990) *Spinal Fusion, Science and Technique,* Springer-Verlag Inc., New York.

DePalma, A.F., Rothman, R.H., Lewinnek, G.E., Canale, S.T. (1972) Anterior interbody fusion for severe cervical disc degeneration. *Surgery, Gynecology & Obstetrics* **134**, 755-58.

Emery, S.E., Bolesta, M.J., Banks, M.A., Jones, P.K. (1994) Robinson anterior cervical fusion. *Spine* **19**(6), 660-63.

Goffin J., Plets C., and Van den Bergh R. (1989) Anterior cervical fusion and osteosynthetic stablilization according to Caspar: A prospective study of 41 patients with fractures and/or dislocations of the cervical spine. *Neurosurgery* **25,** 865-71.

Gore, D.R. and Sepic, S.B. (1986) Anterior cervical fusion for degenerated or protruded discs. *Spine* **9**(7), 667-71.

Karasick D. (1993) Anterior cervical spine fusion: struts, plugs, and plates. *Skeletal Radiol.* **22**, 85-94.

Lunsford, L.D. Bissonette, D.J., Jannetta, P.J., *et al.,* (1980) Anterior surgery for cervical disc disease, parts I and II. *J. Neurosurg.* **53**, 1-19.

McSwain, N.E. Jr., Martinez, J.A., and Timberlake, G.A. (1989) *Cervical Spine Trauma*, Thieme Medical Publishers, Inc. NY.

Newman, M. (1993) The outcome of pseudarthrosis after cervical anterior fusion. *Spine* **18**(16), 2380-82.

Odom, G.L., Finney, W. and Woodhall, B. (1958) Cervical disk lesions. *J.A.M.A.* **166**(1), 23-28.

Randle M.J., Wolf A., Levi L., *et al.,* (1991) The use of anterior plate fixation in acute cervical spine injury. *Surg Neurol.* **36,** 181-89.

Riley, L., Robinson, R.A., Johnson, K.A., Walker, A. E. (1969) The results of anterior interbody fusion of the cervical spine. *J. Neurosurg.* **30**, 127-33.

Ripa D.R., Kowall M.G., Meyer P.R., Rusin J.J. (1991) Series of ninety-two traumatic cervical spine injuries stabilized with anterior ASIF plate fusion technique. *Spine* **16**, 46-55.

Rish, B.L., McFadden, J.T., and Penix, J.O. (1976) Anterior cervical fusion using homologous bone grafts: A comparative study. *Surg. Neurol.* **5**, 119-21.

Robinson, R.A. and Smith, G.W. (1955) Anterolateral cervical disc removal and interbody fusion for cervical disc syndrome. *Bull. Johns Hopkins Hosp.* **96**, 223-24.

Robinson, R.A., Walker, A.E., Ferlic, D.C., Wiecking, D.K. (1962) The results of anterior interbody fusion of the cervical spine. *J. Bone and Joint Surg.* **44A**, 1569-87.

Sedlin, E. and Hirsh, C. (1966) Physical properties of cortical bone. *Acta Orth.* **37**, 29-48.

Seifert V. and Stolke D. (1991) Multisegmental cervical spondylosis: treatment by spondylectomy, microsurgical decompression and osteosythesis. *Neurosurgery* **29**, 498-503.

Smith, G.W. and Robinson, R.A. (1958) The treatment of certain cervical-spine disorders by anterior removal of the intervertebral disc and interbody fusion. *J. Bone and Joint Surg.* **40A**, 607-23.

Suh P.B., Kostuik J.P. and Esses S. I. (1990) Anterior cervical plate fixation with the titanium hollow screw plate system. *Spine* **15**, 1079-82.

Sutterlin, C.E. (1995) Occipitocervical and upper cervical methods of fixation and instrumentation. *Textbook of Spinal Surg.,* J.B. Lippincott Co., Philadelphia.

Sutterlin, C.E., McAfee, P.C., Warden, K.E., Rey, R.M., Farey, I.D. (1988) A biomechanical evaluation of cervical spine stabilization methods in a bovine model: Static and cyclical loading. *Spine* **13** (7), 795-802.

Tew, J.M. and Mayfield, F.H. (1976) Complications of surgery of the anterior cervical spine. *Clin. Neurosurg.* **23**, 424-34.

Tippets R.H., and Apfelbaum R.I. (1988) Anterior cervical fusion with the caspar instrumentation. *Neurosurgery* **88**, 1008-2206.

White, A.A. III, and Hirsh, C. (1972) An experimental study of the immediate load bearing capacity of some commonly used iliac bone grafts. *Acta Orthop. Scandinav.* **42**, 482-90.

White, A.A. III, Jupiter, J., Southwick, W.O., Panjabi, M.M. (1973a) An experimental study of the immediate load bearing capacity of three surgical constructions for anterior spine fusions *Clin. Orthop.* **91**, 21-8.

White, A.A. III, Southwick, W.O., Deponte, R.J., Ganor, J.W., Hardy, R. (1973b) Relief of pain by anterior cervical-spine fusion for spondylosis. *J. Bone and Joint Surg.* **55A** (3), 525-34.

White, A.A. III, and Panjabi, M. M. (1990) *Clinical Biomechanics of the Spine, 2nd Ed.,* J.B. Lippincott Co., Philadelphia.

Whitecloud, T.S. III, and Dunsker, S. (1993) *Anterior Cervical Spine Surgery, Principles and Techniques in Spine Surgery*, Raven Press, NY.

Williams, J.L., Allen, M.B. Jr. and Harkees, J.W. (1968) Late results of cervical discectomy and interbody fusion: Some factors influencing the results. *J. Bone and Joint Surg.* **68** (50A), 277-86.

Witttenberg, R. H., Moeller, J., Shea, M., White, A. A. III, Hayes, W.C., (1990) Compressive strength of autologous and allogenous bone grafts for thoracolumbar and cervical spine fusion. *Spine* **15**(10), 1073-78.

Wood, E. G. III, and Hanley, E. N. Jr. (1992) Types of anterior cervical grafts. *Orthop. Clinics of North America* **23** (3), 475-85.

Young, W.F. and Rosenwasser, R.H. (1993) An early comparative analysis of the use of fibular allograft versus autologous iliac crest graft for interbody fusion after anterior cervical discectomy. *Spine* **18** (9), 1123-24.

Zdeblick, T.A. (1993) Anterior cervical discectomy and fusion. *Operative Techniques in Orthopaedics* **3** (3), 201-6.

Zdeblick T.A., Cooke M.E., Kunz D.N., Wilson D., McCabe R.P. (1994) Anterior cervical disectomy and fusion using a porous hydroxyapatite bone graft substitute. *Spine* **19**, 20.

Zdeblick, T.A. and Ducker, T.B. (1991) The use of freeze-dried allograft bone for anterior cervical fusions. *Spine* **16** (7), 726-29.

Zhang Y., Homsi D., Gates K., *et al.,* (1994) A comprehensive study of physical parameters, biomechanical properties and statistical correlaions of the iliac crest bone wedges used in spinal fusion surgery, IV. Effects of gamma irradiation on mechanical and material properties. *Spine* **19**, 18.

12
Ventilation Tubes

Rodrigo Lambert Orefice
Keith Lobel

INTRODUCTION

Ventilation tubes are widely used in the treatment of middle ear effusion. More than two million tubes are implanted each year in the United States, making them one of the most common types of implant used in the human body. Several materials are used for this implant, in several different geometric designs. Some problems related to the use of ventilation tubes can be attributed to the type of material used and the design.

This chapter provides an overview of published clinical data on different materials and designs of ventilation tubes.

Function, benefits and disadvantages of ventilation tubes

The idea of using hollow tubes, introduced into the tympanic membrane in order to aerate the middle ear, first appeared in 1902, with a procedure using a rubber hollow tube in the eardrum. The results obtained at this time were not good, probably due to primitive sterilization techniques and the lack of surgical microscopes. In 1954 a successful series of operations was reported where synthetic (vinyl) tiny hollow tubes were used to treat otitis media. Since then, ventilation tubes have been frequently used for the treatment of this disease. Tympanotomy tubes, drain tubes, pressure equalization (PE) tubes and 'grommets' are other names for ventilation tubes.

Ventilation tubes are inserted through the tympanic membrane and function by allowing air to pass through it. They enable pressure to equalize between the middle ear and the environment and assist drainage of fluid from the middle ear.

The question of need for a ventilation tube is often controversial. About 70% of children have inflammation or infections in the middle ear at least once during childhood. Loss of aeration in the middle ear, due to the malfunction of the Eustachian tube, will lead to the accumulation of serous or mucoid fluid that, in most cases, will cause infections and diseases, including destruction of ear bones and cholesteatoma. A chronic negative middle ear pressure usually provokes discomfort and hearing loss, which is traumatic for young children. The insertion of ventilation tubes is recommended for:

1. Frequently occurring otitis media: children who suffer recurrent ear infection (more than four per year) that can not be eradicated by the use of antimicrobial treatment;

2. Persistent middle ear fluid, for children who have had fluid for more than 10 weeks, with hearing loss and unsuccessful antimicrobial treatment;

3. Middle ear ventilation disorders which can lead to retraction of the eardrum so that it rests on the incus or stapes bone, causing discomfort and hearing loss.

The rapid benefits brought by the insertion of tubes include: restoration of hearing, relief of discomfort and prevention of structural damage of the middle ear. Long term benefits may include reversal of the mucosal changes of the middle ear, normalization of secretory function and restoration of the ciliary clearance system.

Ventilation tubes may be divided into two groups: *short term tubes* and *long term tubes*. This classification is defined by the average time that a tube should stay in place before being extruded. The time to extrusion depends on several factors such as design of the device, material and surgical procedure. Short term tubes stay in place no more than 18 months, whereas long term tubes are designed to extrude after three years. For two thirds of the children who need ventilation tubes, short term tubes are sufficient, but for the other third a second insertion is necessary, requiring another surgical procedure. For these patients, recurrence of effusion and failure to prevent eardrum atrophy and retraction pockets show that the malfunction of the Eustachian tube persists and a longer term tube is needed. There is another group of patients who require long term tubes, which includes adults with

Eustachian tube obstruction, patients with Down's syndrome, children with craniofacial deformities and patients with tympanic membranes already weakened by prolonged negative middle ear pressure or repeated tube insertions.

Several complications and problems have been reported from the use of ventilation tubes. The most important are described below.

i) Early extrusion. In this case, the tube is extruded before its intended lifetime. The extrusion can be a result of rejection of the inserted material, seen as a prolonged inflammatory reaction with weeping and crusting around the tube, bleeding resulting from granulation, and finally rejection. Early extrusion is usually associated with long-term tubes.

ii) Failure to extrude. In most cases, it is desirable that the ventilation tubes spontaneously extrude from the eardrum after a period of time so as to avoid a second procedure to remove the implant.

iii) Occluded (plugged) tube. For the tube to work, it is necessary that it remain patent. If the lumen becomes occluded by ear wax or dried middle ear fluid, then the tube will not function. In most cases, the occluded tubes can be cleaned by the physician with solvents and appropriate instruments.

iv) Drainage through the tube (otorrhea). This phenomenon is often observed within a few days of insertion of the tube and is part of the inflammatory response to the presence of the material. Drainage usually stops after a few days but prolonged otorrhea may be caused by infection. In that case, antibiotics and other treatments can be used. Prolonged otorrhea can also be produced by an allergy to the implant material.

v) Permanent perforations of the eardrum. Usually after spontaneous extrusion of the ventilation tubes, the hole which remains closes naturally. In other cases, the hole persists and this can lead to hearing loss and morbidity.

vi) Tympanosclerosis. This is a scarring of the tympanic membrane that usually appears as a white or chalky patch and can follow chronic inflammation or trauma.

vii) Infection, hearing loss and cholesteatoma. Cholesteatoma is characterized by formation of cholesterol crystals in the middle ear, after chronic infection which can eventually exert pressure on the tympanic membrane and produce hearing loss.

Table 12.1 summarizes the benefits and complications or problems associated with the use of ventilation tubes.

Table 12.1 Ventilation tubes: benefits and complication or problems

Type of tubes	*Benefits*	*Disadvantages*
Short-term	Restoration of hearing; prevention of structural damage to the ear drum; normalization of secretory function;	Otorrhea; early extrusion; occluded tube; infections;
Long-term	Same as short-term; longer lifetime;	Otorrhea; occluded tube; infections; permanent perforations; tympanosclerosis; sometimes need to be removed;

Surgical procedure

The surgical procedure of tube insertion follows a general sequence, in which the tube is placed in an incision made in the eardrum (myringotomy). The ear canal is cleaned before the tube is inserted and the procedure is almost always done under a microscope. The tube is not sutured to the eardrum, but it remains in position due to the presence of flanges. Each flange stays on one side of the tympanic membrane, providing a method of fixation. Figure 12.1 shows the steps of the procedure.

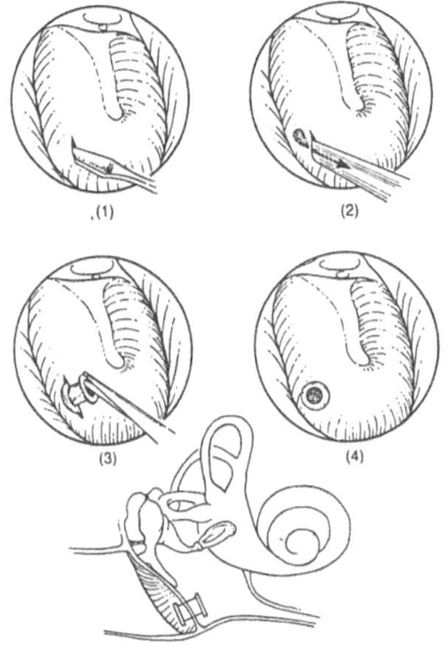

Fig. 12.1 Surgical procedure for insertion of ventilation tubes.

Materials and tube designs

Several materials are used to make ventilation tubes. There have been over 100 different designs developed through trial and error. After studying the different types of implants and the advantages and disadvantages of each, it is possible to define an "ideal" ventilation tube. The ideal ventilation tube can be inserted easily and quickly by the majority of practitioners. It should provide continuous aeration while in position but prevent the entry of water or other material into the middle ear. Design and material composition should not result in long-term pathological changes in the middle ear or tympanic membrane, such as permanent perforations or tympanosclerosis. The implant should stay in place long enough to perform its task and be easily removed if necessary. The desirable time of implantation will vary for each case but generally 18 months is the minimum expected lifetime.

In the past three decades, a large number of ventilation tube designs have been proposed. The different designs are usually associated with the developer's name. Most of the ventilation tubes have three parts: a medial end, the body and the lateral end (see Fig. 12.2).

Medial end. The form of the middle ear end is crucial as it determines the ease of insertion or removal and the rate of extrusion. Therefore, most

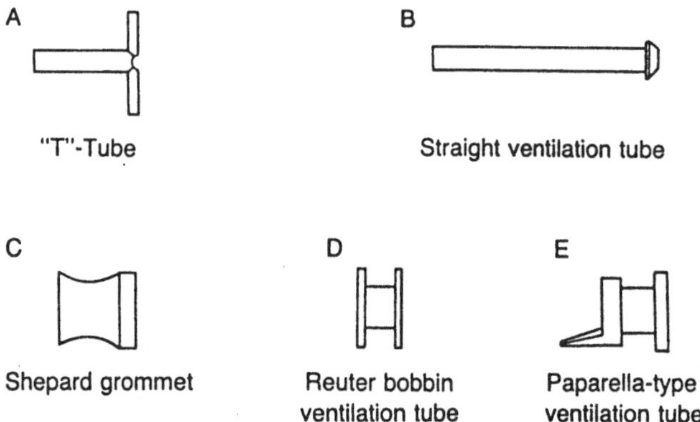

Fig. 12.2 Examples of ventilation tubes.

modifications involve this part of the device. The most common design is a flange with either circular, elliptical or notched form. The size (height) of the flange is a crucial factor in determining the rate of extrusion. For long-term implants, a very wide flange is used. The angle between the flange and the body is also an important feature, since it determines the degree of difficulty of insertion and also effects extrusion rate.

Body. All tube bodies are circular in form and contain a central lumen. The external diameter of the body is often less than 2.5 mm. The internal diameter is also important. For long term implants, a larger diameter is preferred to avoid occlusion, but it should be small enough to prevent the entry of water and foreign bodies. The dimension for this inner diameter ranges from 1 to 1.5 mm. This dimension may vary from material to material, since the tendency of water to go into the hollow tube is a function of the surface tension of the water/material interface. The difference in pressure (ΔP) between the inside and outside of the tube that drives the liquid through the orifice is expressed by $\Delta P=2\gamma/r$, where γ is the surface tension between water and the tube material. The length of the body also varies. Shorter sizes are preferred to reduce the occurrence of occlusion.

Lateral end. This end is usually in the form of a flange. In some designs, this portion of the device is manufactured with a wire attached to facilitate removal.

Figure 12.2 shows examples of ventilation tubes commonly used. All try to optimize the conditions described above. The 'T' tube shown in this figure is an example of a long-term device. Other devices can be rotated after insertion to lock in place, or flap open like an umbrella when placed in the eardrum. Most of them are colored to facilitate easy identification by the physician.

Several materials have been used in the past, but only a few have extensive usage. More than 90% of ventilation tubes used today are made from either polyethylene, polytetrafluoroethylene (Teflon®) or silicone rubber. Other materials used less frequently are stainless steel, titanium, hydroxyapatite and gold. Most of these materials are still in the research stage. Hydroxyapatite and titanium are the materials tested most recently as ventilation tubes. As a result, there is little clinical data yet available for them. Most of the reported clinical data on ventilation tubes involve the three polymeric materials cited earlier. Most of the designs shown in Fig. 12.2 are made from one of these materials.

Table 12.2 tabulates tube designs available commercially and described in the literature, the material used and other characteristics of the device.

DEFINITION OF SUCCESS

The properties of the 'ideal' implant have been defined. From the published data, it is possible to identify the characteristics that are frequently related to problems. Early extrusion and permanent perforations are discussed in this chapter, since they are currently the most common problems. Longer implantation times with reduced incidence of permanent perforation are desirable.

Both problems have been treated by researchers and professionals of the field by change in design. In addition to design, the material used and the method of insertion are important factors affecting extrusion rates, and permanent perforations are usually associated with long term implantation.

In Table 12.2, it can be seen that comparison between materials is difficult since differences in materials and designs cannot be separated. Only by examining identical designs in different materials (and vice-versa) can the effect of the material or design alone be established.

Table 12.2 Materials and design of some of the most common ventilation tubes

Ventilation tube	*Material*	*Design*	*Implantation time*
Shepard	PTFE	double flanges	short term
Exmoor	Polyethylene	double flanges	short term
Goode 'T'- tube	silicone rubber	two limbs at right angles on the middle ear end	long term
Bobbin	316L stainless steel	bobbin-type with flanges	short/long term
Gold-plate	gold	not described	short term
Per-Lee	silicone rubber	large flanges - angle between body and flanges = 60-70°	long term
Biointegrated hydroxyapatite (HA)	Dense HA	eccentric disk-like flange in the middle ear end	short term
Shah	Polyethylene	circular flange	short/long term
Armstrong	PTFE	two small limbs at right angles on the middle ear end	short term
Titanium	Titanium	double flanges	short term
Collar-button	PTFE	wider flanges at right angles	short term

Although the data evaluated are not optimal in design, statistical analysis will be used to detect effect of the material on the problems described. All designs that use the same material are grouped for analysis.

DATA ANALYSIS

Twenty-three papers were evaluated to produce the clinical data on success rates of ventilation tubes (see the Reference List).

Figures 12.3 and 12.4 show the results of the comparison among the several tube designs related to time of extrusion. Large variations in performance are obtained with different designs and materials. Figure 12.4 shows the extrusion rates for long-term devices with large flanges and diameters.

A Student's t-test was done on each curve shown in Figs. 12.3 and 12.4 to test the hypothesis that any one specific design could lead to lower extrusion rates. To apply the Student's t-test, an approximated normal distribution is needed. By normalizing the axes of Figs. 12.3 and 12.4 in terms of fraction of the available devices extruded at a specific time, it was possible to obtain normal-type curves. For a probability of 95%, all the curves' results stayed in the tails of the Student curve, supporting the hypothesis that different combinations of material and design can lead to very different performances.

Figure 12.5 represents the other problem, perforation, used as an indicator of success of a ventilation tube. In this figure, the different designs are compared for permanent perforation. An important result deduced from this figure is that although it is possible to achieve longer periods without extrusion, these longer periods are often achieved at the expense of permanent perforations. There is no ventilation tube today that can satisfy the need for a long implantation time without extrusion or permanent perforation.

In Figs. 12.6 and 12.7, the extrusion rates of ventilation tubes are compared as a function of material only. Data from tubes made of the same material, but of different designs are combined. Some of the materials shown in Figs. 12.6 and 12.7 have a much smaller number of samples than others. For example, hydroxyapatite and titanium have fewer clinical results than PTFE (Teflon®) and PE (polyethylene).

By use of the Student's t-test, it was possible to compare the significance of the results ($p = 0.05$) and the methodology applied (sum of clinical results from different devices made from the same material). The results of the test show no significance, i.e., it is not possible to say, from the data collected, that the material used affects the performance of a ventilation tube to a significant degree.

In terms of the materials used in the fabrication of the tubes, 90% of them are made from polymers, PE, PTFE or silicone rubber. Since these materials are all relatively inert, no large difference in behavior should be expected. However, these materials differ in stiffness and silicone rubber

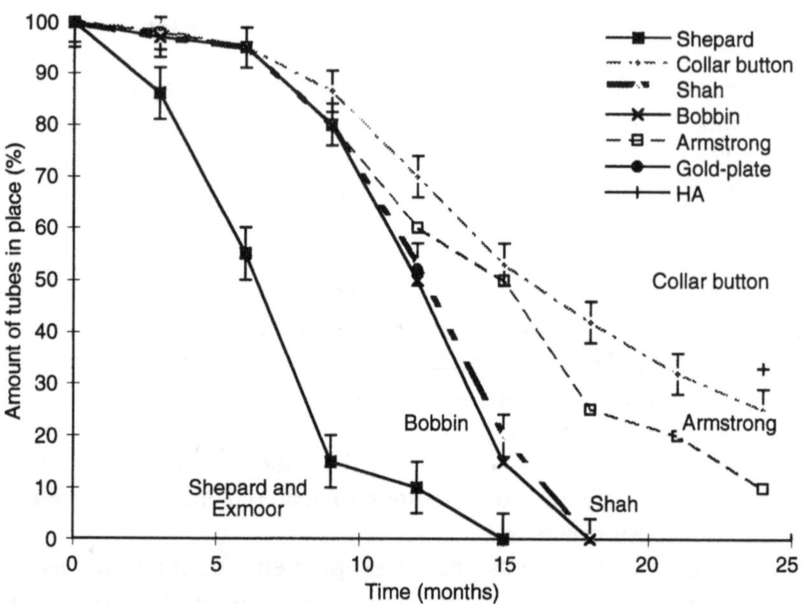

Fig. 12.3 Extrusion rates of several short-term ventilation tubes.

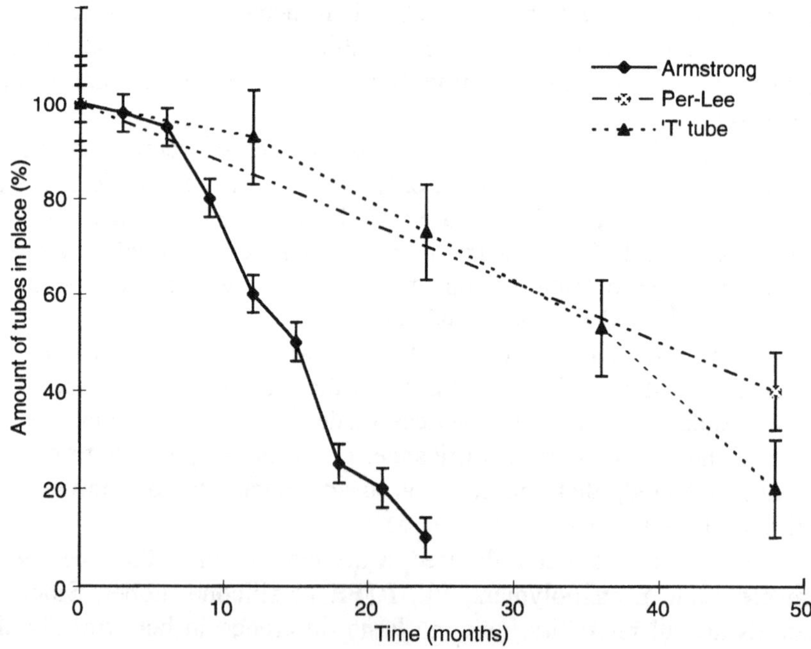

Fig. 12.4 Extrusion rates of long-term ventilation tubes.

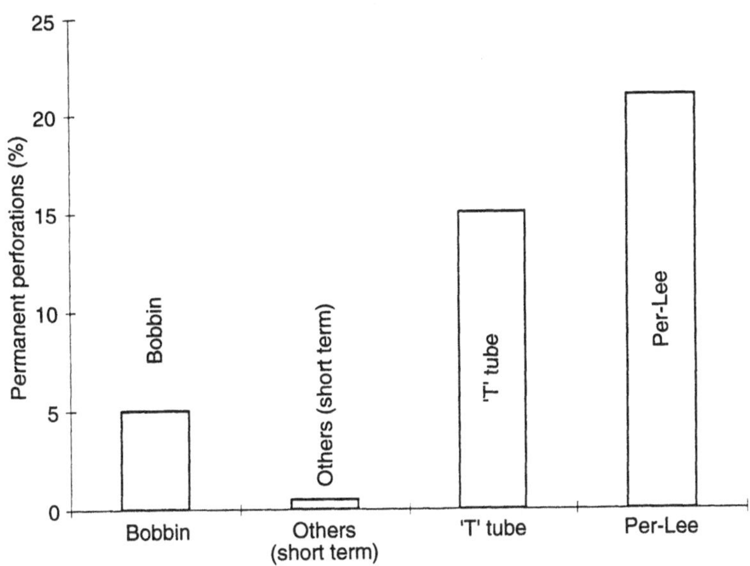

Fig. 12.5 Percentage of permanent perforations associated with different ventilation tubes.

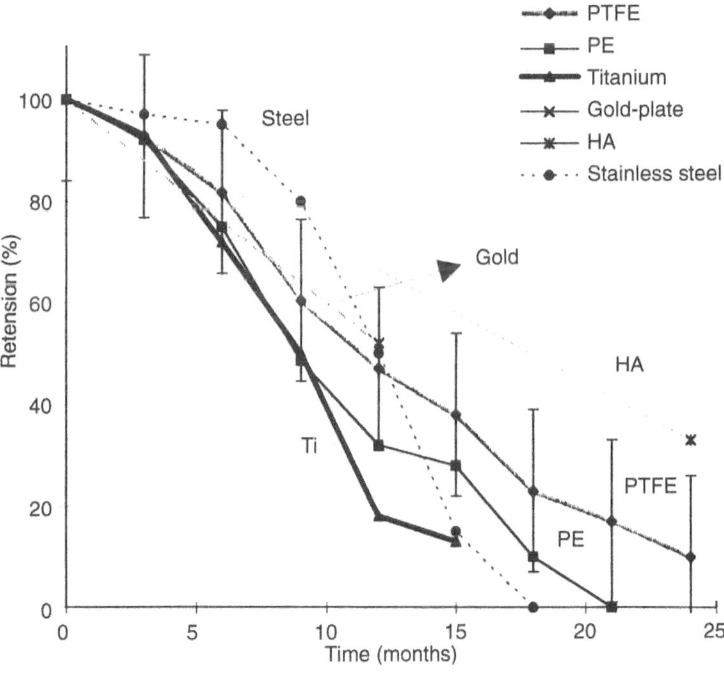

Fig. 12.6 Extrusion rates of several materials used to produce short-term ventilation tubes.

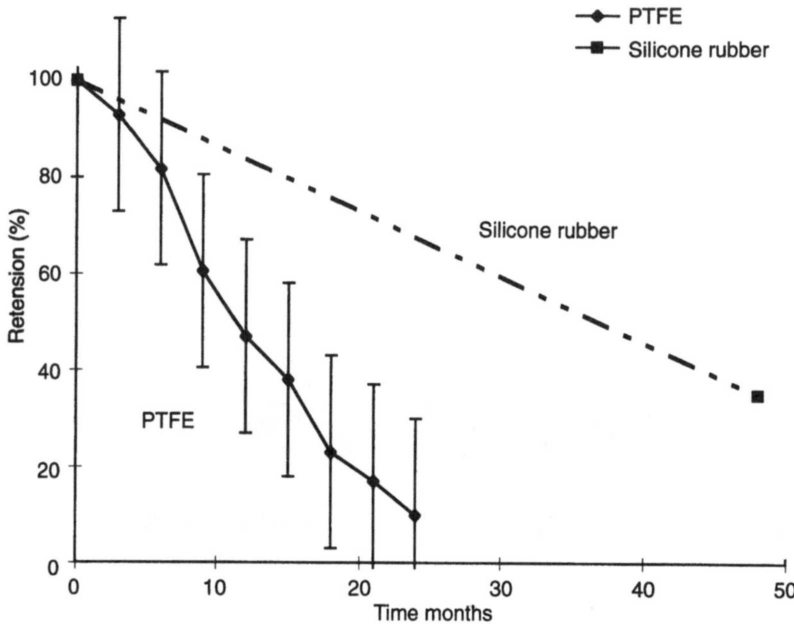

Fig. 12.7 Extrusion rates of two materials used to produce long-term ventilation tubes.

ventilation tubes (Fig. 12.7) are the most stable due to the ease with which this material can be deformed and passed through the incision made in the eardrum, after which it returns to its original shape, producing compressive forces in the eardrum which hold it in place and minimize movement.

Some of the causes of early extrusion in ventilation tubes are related to so-called rejection by the body, prolonged inflammation and infection. These problems can be related to the type of material used to fabricate the tube. The main concern of the tube designers and users was to choose a material that would provoke minimal tissue reaction. The polymeric materials chosen (polyethylene, silicone rubber, PTFE) are sometimes called "bioinert", but they always lead to the formation of a thin fibrous capsule that surrounds the implant. These materials do not adhere to the tissue and movement at the interface in use can lead to tissue damage, infection and extrusion. Use of a "bioinert" material cannot prevent this reaction although good technique can minimize it. Alternatively, the implantation of a bioactive implant is expected to lead to better results by lowering extrusion rates, since these materials, by adhering to the surrounding tissue, produce a stable interface that can resist movement and thus reduce inflammation, infection, and extrusion. Figure 12.6 shows that HA, a bioactive ceramic used for this purpose, appears to have an advantage, however there is not yet sufficient data to confirm this nor is the mechanism clear since HA behaves as an inert material in soft tissue.

CONCLUSION

There is no ventilation tube today that provides long term implantation without extrusion or permanent preformation of the tympanic membrane. The clinical results investigated in this chapter provided little information regarding the effect of the material on the extrusion rate and production of permanent perforation by the devices. This was mainly due to the differences in design between devices of similar materials. Therefore, the influence of the material on the performance of the tube cannot be separated from the influence of design.

Most of the materials used today to fabricate ventilation tubes are bioinert. This bioinert behavior is presumed to be responsible for the high rates of extrusion shown in this work. Use of a bioactive material may minimize this cause of extrusion, and allow simplification of designs which may be contributing to complications such as permanent perforations.

REFERENCES

Abdullah, V.A., Pringle, M.B. and Shah, N.S. (1994) Use of the trimmed Shah permavent tube in the management of glue ear. *J. Laryngol. Otol.,* **108** (4), 303-6.

Crysdale, W.S. (1976) Comparative study of various ventilating tubes. *Ann. Otol. Rhinol Laryngol.,* **85** (3-4), 268-69.

Gibb, A.G. (1986) Long-term tympanic ventilation by Per-Lee tube. *J. Laryngol. Otol.,* **100** (5), 503-8.

Gibb, A.G. and Mackenzie, I.J. (1985) The extrusion rate of grommets. *Otolaryngol. Head Neck Surg,* **93** (12), 695-699.

Grundfast, K. and Carney, C. (1987) *Ear Infection in Your Child,* XII, Compact Books, Publishers, Hollywood, FL, p. 283.

Hawthorne, M.R. and Parker, A.J. (1988) Perforations of the tympanic membrane following the use of Goode-Type 'long term' tympanostomy tubes. *J. Laryn. Otol.* **102** (11), 997-99.

Hughes, L.A. and Wright I.D. (1988) Tympanostomy Tubes: Long-term effects. *Am. Fam. Physician,* **38** (5), 186-90.

Hussain, S.S.M. (1992) Extrusion rate of Shah and Shepard ventilation tubes in children. *J. Ear Nose Throat* **71** (6), 273-75.

Jahn, A. F. (1993) Middle ear ventilation with Hydroxylvent tube: review of the initial series. *Otolaryngol. Head Neck Surg.,* **108** (6), 701-5.

Jahn, A.F. (1991) A biointegrated hydroxylapatite ventilation tube for definitive treatment of chronic Eustachian tube obstruction. *Otolaryngol. Head Neck Surgery,* **105** (11), 757-60.

Karlan, M.S., Skobel, B., Grizzard, M., *et al.* (1980) Myringotomy tube materials: bacterial adhesion and infection. *Otolaryngol. Head Neck Surg,* **88** (6), 783-94.

Levine, S., Daly, K. and Giebink, G.S. (1994) Tympanic membrane perforations and tympanostomy tubes. *Ann. Otol. Rhinol. Laryngol. Suppl.,* **163** (5), 27.

Mackenzie, I.J. (1984) Factors affecting the extrusion rates of ventilating tubes. *J. Royal Soc. Medicine,* **77** (9), 751-53.

Meyerhoff, W.L. (1981) Use of tympanostomy tubes in otitis media. *Ann. Otol.,* **90**, 537-41.

Moore, P.J. (1990) Ventilation tube duration versus design. *Ann. Otol. Rhinol. Laryngol.,* **99**, 722-23.

Reuter, S.H. (1968) The stainless steel bobbin middle ear ventilation tube. *Trans. Am. Acad. Ophthalmol. Otolaryngol.,* **72** (1), 121-2.

Shone, G.R. and Griffith, I.P. (1990) Titanium grommets: a trial to assess function and extrusion rates. *J. Laryngol. Otol.* **104** (3), 197-99.

Soderberg, O. and Hellstrom, S. (1987) Consequences of using hyaluronan-coated tympanostomy tubes. *Acta Otolaryngol.,* **442**, 50-3.

Soderberg, O. Hellstrom, S. (1987) Effects of different tympanostomy tubes (Teflon and stainless steel) on the tympanic membrane structure. *Recent advances in otitis media. Proceedings of the Fourth symposium.* Bal. Harbour, FL.

Tami, A., Kennedy, K.S. and Harley, E. (1987) A clinical evaluation of gold-plated tubes for middle-ear ventilation. *Arch. Otolaryngol. Head Neck Surg,* **113** (9), 979-80.

Watson, C. and Mangat, K.S. (1988) A comparison of audiometric performance and complications of T tubes and Shepard grommets. *J. Laryngol. Otol.,* **102** (8), 677-79.

Weigel, M.T., Parker, M.Y.; Goldsmith, M., *et al.* (1989) A prospective randomized study of four commonly used tympanostomy tubes. *Laryngoscope,* **99** (3), 252-56.

Wielinga, E.W.J. and Smyth, G.D.L. (1990) Comparison of the Good T-tube with the Armstrong tube in children with chronic otitis media with effusion. *J. Laryngol. Otol.* **104** (8), 608-10.

13

Ossicular Replacement Prostheses

Keith D. Lobel

INTRODUCTION

The field of biomaterials has, to a large extent, derived most of its common materials from other fields of materials science. Similarly, biomaterials in otolaryngology have been previously utilized in other biomedical arenas. As with most implants, homografts or autografts are generally considered the 'gold standard' against which all other materials are compared. But lack of availability, among other reasons, prevents homografts from being the most common material used for replacement of ossicles. For reasons discussed later, ossicular replacements are usually made of synthetic materials such as polyethylene and ceramics.

This chapter evaluates the clinical success of various types of synthetic implants used for reconstruction of the ossicular chain. Data are accumulated from more than 40 papers published within the last decade, and statistical analyses have been performed to determine significant differences in success rates, which have been defined in terms of extrusion and/or displacement of the implant, and closure of the air-bone gap (A-B gap)*. Indeed 'extrusion and persistent or recurrent conductive hearing loss are the most common causes of operation failure' (Emmett *et al.,* 1986). Interestingly, Shea and Emmett (1984) cite the following factors, in order of decreasing importance, as being most critical in achieving clinical success: 1) pathology present in the area to be implanted and the surgical skill

*The A-B gap is the difference between the preoperative bone conduction level and postoperative air conduction level, where the air conduction sensitivity is typically monitored using earphones and the bone conduction sensitivity is measured by placing a sound probe on the auditory mastoid prominence, by-passing the middle ear components, including the ossicles. Both measure the smallest detectable sound level at various frequencies (typically 500, 1000, and 2000 Hz).

employed in the implantation, 2) biomechanical properties of the implant, and 3) biocompatibility of the material implanted.

BACKGROUND

There are many reasons for surgical restoration of the ossicular chain (Reck and Helms, 1984; Jackson, 1983). *Cholesteatoma* involves the formation of cholesterol crystals in the middle ear, which can accumulate and exert pressure on the surrounding tissue, including the ossicles, leading to damage of the bones. *Otitis media* is characterized by inflammation of the middle ear, often from bacterial or viral infection which can attack the ossicles resulting in conductive deafness. This can also spread to the mastoid bone of the inner ear which contains the air cells, resulting in *mastoiditis*. *Otosclerosis* is a hereditary disorder causing deafness in adult life due to an overgrowth of the mastoid bone, separating it from the middle ear and preventing sound conduction. Many patients exhibit a combination of diseases (i.e., chronic otitis media with effusion) leading to more severe symptoms and requiring more drastic measures of treatment.

Common goals of middle ear surgery involving the ossicles, then, include long-term aeration of the middle ear, (Smyth, 1982; Silverstein *et al.*, 1986) restoration of conductive hearing, (Jackson *et al.*, 1983; Silverstein *et al.*, 1986) repair of the tympanic membrane to achieve normal compliance, (Smyth, 1982; Silverstein *et al.*, 1986) and infection control of the Eustachian tube and restoration of its function (Jackson *et al.*, 1983).

Ossicular chain reconstructions have been performed since 1875, but the first synthetic implant was not introduced until 1952. Before this time, various forms of natural materials were used for reconstruction of the ossicular chain including autologous bone, autologous ossicles, autologous cartilage, allogenic cartilage, and bone grafts (Reck, 1984).

While homografts and autografts have been considered as the best available material, (Austin, 1984; Grote, 1986; Podoshin *et al.*, 1988; Merwin, 1986; Reck *et al.*, 1988; Toner *et al.*, 1991; Epstein and Sataloff, 1986; Mangham and Lindeman, 1990) their use has several disadvantages. Often a prolonged operative time is required to shape the part (Nikolaou *et al.*, 1992). Partial resorption and atrophy of the homografts or autografts may occur, and bony fixation to the surrounding tissue following lateralization is a potential problem (Grote, 1986; Jahnke *et al.*, 1983).

Transmission of diseases such as AIDS require special handling in addition to the established steps of harvesting, preparation, and storage (with limited shelf life), (Goldenberg, 1990, 1992) all of which can be avoided through the use of alloplastic materials. Finally, various cultures prohibit the use of homologous materials for middle ear implants due to religious beliefs (Podoshin *et al.*, 1988).

In an attempt to circumvent these problems, many different synthetic materials have been used since initial introduction of pallavite. These are shown chronologically in Fig. 13.1. Of these, only a few have been used in significant numbers: non-porous polyethylene, Teflon®, Proplast®, Plastipore®, alumina, hydroxyapatite (HA), and Ceravital® (Brackmann *et al.*, 1984; Hughes, 1987; Colletti *et al.*, 1987; Grote, 1987; Reck, 1984; Jahnke *et al.*, 1983).

Non-porous polyethylene was first used by Shea in 1958 for stapedial surgery (Shea, 1958). However, traumatic effects in the surrounding tissue as well as marginal hearing gains led to exploration of other materials (Reck, 1984). Teflon® generally gave better hearing results with less trauma to the surrounding tissue, presumably by becoming ensheathed by mucosa in the middle ear. Teflon® is still used today in combination with wire for stapedectomies, although neither it nor polyethylene have been accepted for tympanoplasty procedures due to resorption and spontaneous rejection by the tympanic membrane (Reck, 1984).

In 1976 two different porous plastic materials were introduced as potential materials for ossicular replacement: Proplast® and Plastipore®. Proplast® is a combination of polytetrafluoroethylene (Teflon®) and vitreous carbon, and contains pores between 100 and 500 μm accounting for 70 to 90% of its volume (Shea and Homsy, 1974). It is claimed that the pores permit 'precipitation of proteins within the pores in a relatively undenatured form that camouflage the implant from the body's immunologic rejection mechanism' (Shea and Emmett, 1984). Adverse reaction with the body in other medical applications has led to discontinued use of Proplast® for biomedical applications, including ossicular replacements.

®Teflon: Dupont, Wilmington, Delaware

®Proplast: Vitek, Inc., Hazelwood, MO

®Plastipore: Richards Mfg., Memphis, TN

®Ceravital: E. Leitz Wetzlar GmbH, D-6300 Wetzlar, Germany

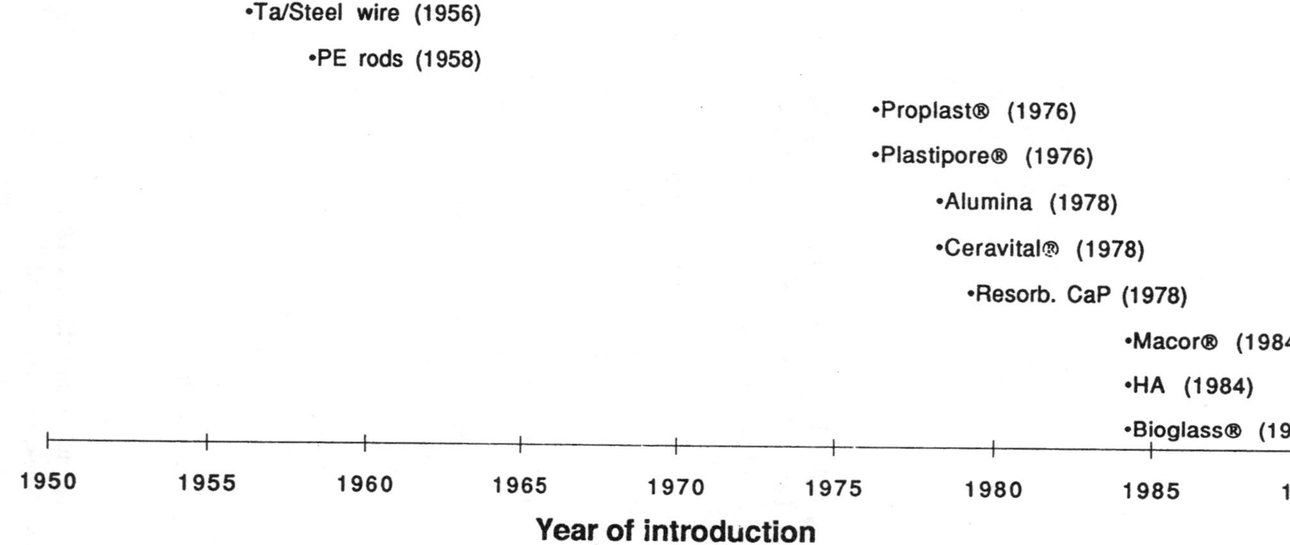

Fig. 13.1 First use of various synthetic materials as ossicular chain replacements.

Plastipore® is a porous 'sponge-like' high-density polyethylene (Reck, 1984). It is said to be more easily handled in the operating room, but exhibits higher extrusion rates than Proplast® (Epstein and Sataloff, 1986). The porosities of Proplast® and Plastipore® allow tissue ingrowth, which fixes and stabilizes the implant. It is generally accepted that placement of cartilage (or bone paté) between the implant and the tympanic membrane can significantly reduce extrusion rates (Babighian, 1984; Jahnke *et al.*, 1983; Emmett *et al.*, 1986; Brackmann and Sheehy, 1979; Shea and Emmett, 1984; Sanna *et al.*, 1984b; Portmann, 1984). Gamoletti *et al.* (1984) concluded that 'no histologic feature supports a biologic cause of extrusion, and that extrusion instead is related to biofunctional characteristics.' Palva and Makinen (1983) point out that, while cartilage interposition may reduce extrusion rates, this may only be a temporary finding, since past studies using cartilage in tympanoplasty have reported resorption or softening of the material. The surgical technique of sizing the implant in the operating room is of critical importance, as this step determines the amount of pressure exerted on the tympanic membrane by the malleus or implant, and therefore determines to a large extent the rate of extrusion.

While results were much better with the biocompatible porous polymer implants, there was still much room for improvement, particularly with respect to extrusion rate (Grote, 1986; Niparko *et al.*, 1988; Sanna *et al.*, 1985). Another class of biomaterials used for ossicular replacement that is gaining acceptance is bioactive ceramics. As shown in Fig. 13.1, bioactive ceramics account for most of the new materials used in otology since Proplast® and Plastipore® were introduced.

Bioactive ceramics (and glasses) exhibit chemical bonding at the biomaterial-tissue interface. This unique mechanism of fixation is achieved by the formation of chemical bonds between the implant and host tissue (Hench and Wilson, 1993). The bonding is achieved by controlled reactivity at the implant surface, including dissolution of multivalent ions and precipitation of a hydroxyapatite layer. The hydroxyapatite provides sites for collagen deposition and integration, resulting in strong chemical fixation of the implant. Otologic animal studies using Ceravital® began in 1981 by Reck (Blayney *et al.*, 1986) and Bioglass® in 1982 by Merwin *et al.*

®Bioglass: University of Florida, Gainesville, Florida 32610

(1982). Excellent tolerance led to further investigation of bioactive materials for middle ear implants.

Based upon these extensive animal studies it was concluded that the ability to bond with the surrounding tissue would stabilize the ossicular implant, thereby reducing extrusion rates. In addition, the inherent binding of the material eliminates the need for the interposition of cartilage between the tympanic membrane and the implant (Reck and Helms, 1985). Most of the literature reviewed in bioactive ceramics concentrates on either hydroxyapatite or the glass-ceramic Ceravital®. Hydroxyapatite is a well-known calcium-phosphate ceramic used for various biomedical implants (Hench and Wilson, 1993) and is particularly useful for middle ear implants in that it can be shaped in individual sizes easily, and does not require load-bearing capacity (Grote, 1987). Similarly, numerous studies have been conducted using Ceravital®, glass-ceramic, as ossicular bone replacements (Reck, 1984; Reck and Helms, 1984; Reck *et al.*, 1988; Blayney *et al.*, 1986). Bioglass® has been shown to exhibit even lower extrusion rates than Ceravital® after long implantation periods, and is more easily shaped in the operating room (Merwin, 1986; Merwin *et al.*, 1984; Wilson *et al.*, 1985). Table 13.1 compares the chemical composition, physical properties, and

Table 13.1 Composition and properties of bioactive glasses and ceramics used for ossicular implants (Hench and Wilson, 1993, Reck, 1984)

Substance	*Bioglass® (45S5)*	*Ceravital®*	*Hydroxyapatite*
SiO_2	45.0 w/o	40-50 w/o	
Na_2O	24.5	5-10	
CaO	24.5	30-35	
P_2O_5	6.0	10-15	$Ca_{10}(PO_4)_6(OH)_2$
MgO		2.5-5	
K_2O		0.5-3	
Property			
Elastic Mod. (GPa)	35	100-150	80-110
Comp. Strength (MPa)	42	500	500-1000
Bend. Strength (MPa)			115-200

bioactivity index of the three types of bioactive implants used for ossicular replacement. An important difference is the low elastic modulus (stiffness) of Bioglass® (see Chapter 1) and low hardness which makes it very easy to contour into shapes in the operating theater (Merwin *et al.*, 1984).

In the quest for the optimum biomaterial for ossicular implants, it must first be realized that no material will fit the requirements of all situations (Blayney *et al.*, 1992; Jahnke *et al.*, 1983). However, there are several general requirements that can be identified for otological applications. As is true for any biomaterial, it must be biocompatible and exhibit good histological responses, be readily available and be easy to use (Jahnke *et al.*, 1983; Goldenberg, 1990; Shea and Emmett, 1984). Important mechanical properties include shapability (either by molding or machining), effective sound energy transmission, rigidity/stiffness, and minimal load-carrying capability (Podoshin, 1988; Brackmann *et al.*, 1984). It should also exhibit long-term stability and not resorb or cause atrophy of the surrounding tissue (Jahnke *et al.*, 1983; Brackmann *et al.*, 1984).

METHODS

Data has been reviewed and analyzed from over 40 papers published in the last decade. In the context of a statistical literature review, it was decided that success would be evaluated in terms of two characteristics: extrusion (and displacement) rate, and closure of the air-bone gap (A-B gap). These criteria permitted a reasonable number of studies using similar implant systems and surgical methods to be combined for statistical analysis. Particular attention was given to the implant material, type of implant (i.e., partial ossicular replacement (P) or total ossicular replacement (T)), number of cases reported, interposition of cartilage, or not, and period of follow-up evaluation.

Factors not considered in the statistical analyses include implants made of natural biomaterials (homografts, autografts, or cartilage), stapedectomy procedures, influence of middle ear disease, staging of procedures, histological responses, patient age, date of implantation, and less common modes of failure. These factors have been considered elsewhere (Giddings and House, 1992; Sanna *et al.*, 1984a; Shea and Emmett, 1984; Frootko, 1984; Reck *et al.*, 1988; Grote, 1990; Smyth, 1982; Silverstein *et al.*, 1986; Gersdorff *et al.*, 1989; Blitterswijk *et al.*, 1990; Palva and Makinen, 1983;

Gamoletti *et al.*, 1984; Mercandino and Tarasido, 1975; Colletti *et al.*, 1987; Yamamoto and Iwanago, 1986).

By far the limiting factor in comparing results of various studies is the number of variables in middle ear procedures, and therefore the number of available methods and forms of data presentation. Three primary classifications considered here are 1) the material used, 2) the type of implant used (P) vs. (T), 3) and the presence or absence of cartilage interposed between the implant and tympanic membrane. This leads to four possible implant systems for each type of material. Data included herein consider six different materials: porous polyethylene (Plastipore®), hydroxyapatite, Ceravital®, Al_2O_3 (with fascial graft), Bioglass®, and a carbon-carbon composite. Data for most of these materials were subdivided into total ossicular replacement implants (Ts) and partial ossicular replacement implants (Ps), depending on which of the three ossicular bones were being replaced. The T classification is the more basic of the two, referring to simultaneous replacement of all three ossicles (malleus, incus, and stapes) through the use of a single implant. Based on the data presented in the articles reviewed, Ps herein refer to: 1) replacement of only the incus, 2) replacement of the incus and malleus, or 3) replacement of incus and stapes. Note that this excludes several other replacements that could be classified as Ps such as stapedectomies and malleus replacements.

Figure 13.2 shows schematics (taken in part from Reck and Helms, 1985) of the P and T classifications used for this study. The purpose of cartilage interposition is primarily to decrease the incidence of extrusion or dislocation (Yamamoto and Iwanaga, 1986). Data from studies using fascia grafts or tissue glue are combined with studies using cartilage interposition, as they are employed in the same manner for the same purpose.

Extrusion. Extrusion rates were evaluated for each system as a function of time after implantation. Also included in these data were displaced prostheses since both extruded and displaced prostheses have, in part, a common cause of failure: lack of fixation. It should be noted, however, that typically extrusion is also indicative of prostheses exerting pressure on the tympanic membrane, a phenomenon not necessarily associated with displacement. Data points from the same implant system at the same postopetative time period were combined by adding the number of implants extruded in each study to the total number of prostheses implanted. Categorical data was analyzed using Fischer's Exact test, since the unpaired

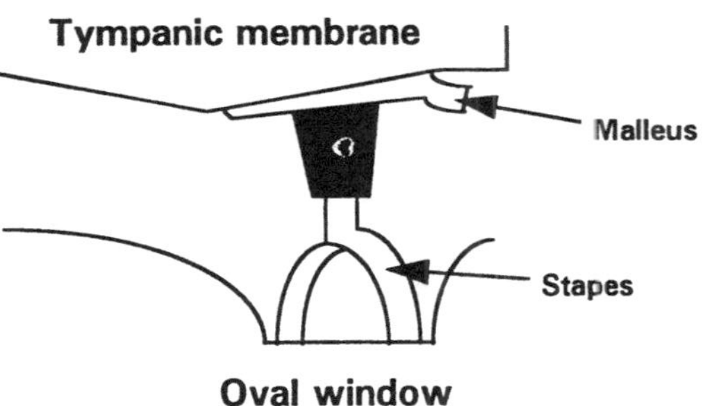

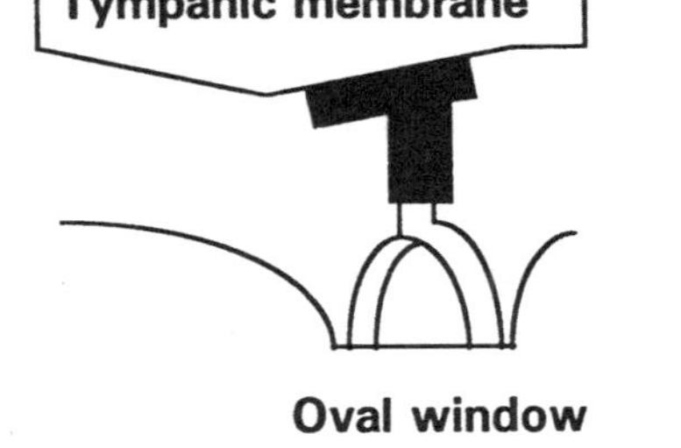

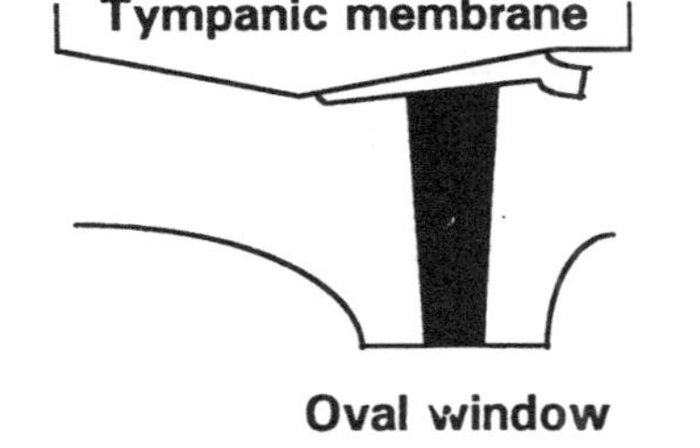

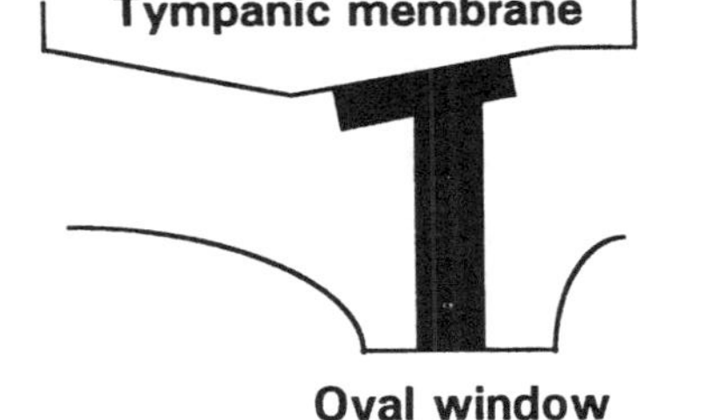

Fig. 13.2 Schematics of T (total) and P partial systems.

data represent true contingency tables. Exact two-sided P-values are reported. P-values less than 0.05 indicate a significant difference between the categories compared.

Air-Bone Gap. A-B gap distributions were evaluated for each system. This was done in the same manner as the extrusion data, by adding the number of patients with the same A-B gap to the total number of patients studied. A linear regression analysis was performed on each system. Data series having a squared correlation coefficient lower than 0.90 were excluded from further analysis. Following the literature, success was defined by an A-B gap of ≤ 15 dB for Ts and ≤ 25 dB for Ps (Nikolaou *et al.*, 1992). Success rates for each system at the A-B gap of interest were calculated from the linear regression fit. Categorical data was analyzed using Fischer's Exact test. Exact two-sided P-values are reported.

RESULTS AND DISCUSSION

Extrusion rate

Figure 13.3 shows the extrusion rates over time for each of the porous polyethylene (Plastipore®) systems. Undoubtedly, the most significant data from such an analysis are extrusion rates at very long times. Unfortunately, these are the data points for which there is very little data available. The number of samples for each curve at the 5-year mark is given. While the numbers are much higher at earlier postoperative times (often several hundred patients), no significant trends are found before 4-years. After 4 years there are signs of dispersion between the four systems, but more data at longer times are required to validate this observation.

Table 13.2 shows the results of a statistical analysis of the extrusion data at the 5-year mark. Of the various sets of classifications compared, four showed significant differences:

1. All PE Ps vs. All PE Ts (extremely significant),
2. All PE implants using cartilage vs. all PE implants not using cartilage,
3. PE Ts vs. PE Ps both without cartilage,
4. Bioactive ceramics vs. all PE Ts and Ps.

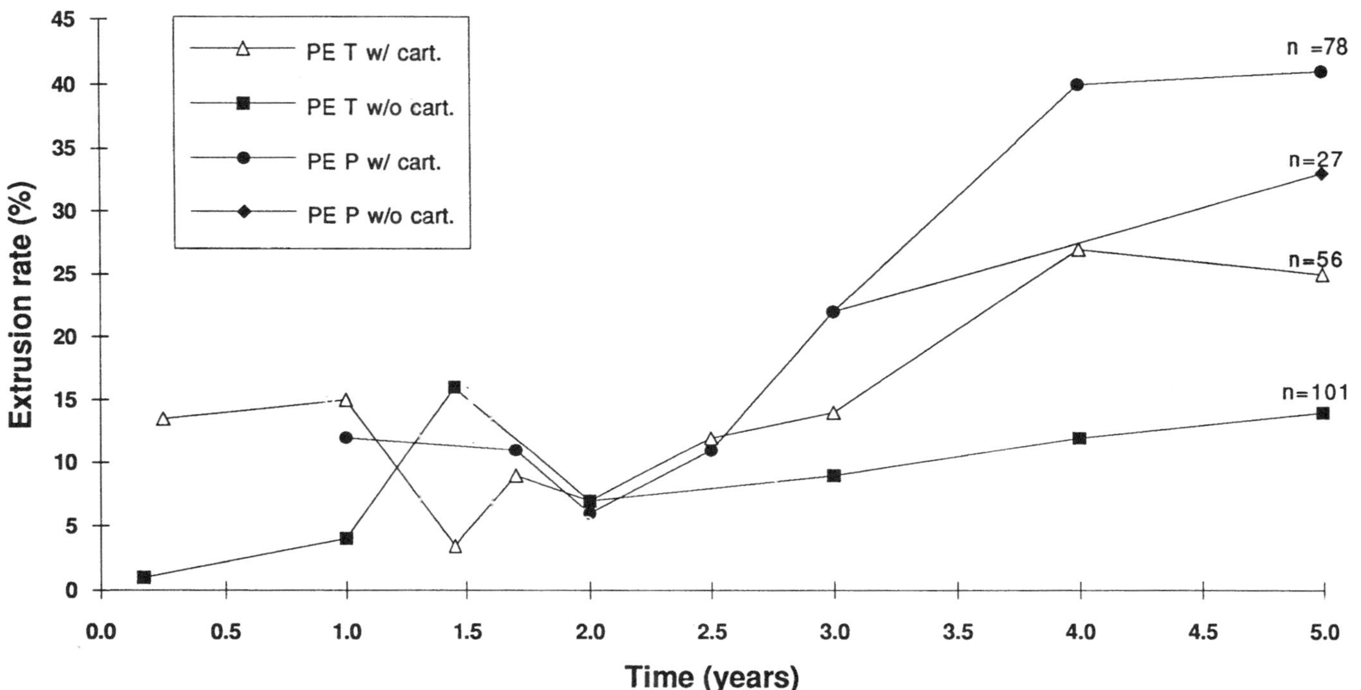

Fig. 13.3 Extrusion rate vs. time of implantation of various MEP materials.

Table 13.2 PE-System comparisons of extrusion data at 5 years after surgery to identify significant differences (via Fischers exact test)

	Success (%)	*Failure (%)*	*Statistical Significance*
All PE Ps	39.0	61.0	
All PE Ts	17.8	82.2	*P = 0.00002
All PEs w/ cart.	34.3	65.7	
All PEs w/o cart.	18.0	82.0	*P = 0.0031
PE Ts w/ cart.	25.0	75.0	
PE Ts w/o cart.	13.9	86.1	P = 0.8700
PE Ps w/ cart.	41.0	59.0	
PE Ps w/o cart.	33.3	66.7	P = 0.6477
PE Ts w/ cart.	25.0	75.0	
PE Ps w/ cart.	41.0	59.0	P = 0.0659
PE Ts w/o cart.	13.9	86.1	
PE Ps w/o cart.	33.3	66.7	*P = 0.0261
Bioactive ceramics (5+ yrs.)	92.3	7.7	
PE Ts w/o cart.	86.1	13.9	P = 0.0529
Bioactive ceramics (5+ yrs.)	92.3	7.7	
All PE Ts and Ps	64.2	35.8	*P < 0.0001
Bioglass® Ts and Ps (5+ yrs.)	96.7	3.3	
Ceravital® Ts	91.0	9.0	P = 0.2006

Other than the third significant comparison, no significant differences were found among the PE prostheses alone. The favorable results of the PE Ps compared with PE Ts were expected, as others have found similar results. Interestingly, the second significant comparison given above concludes that implants <u>without</u> cartilage interposition have significantly lower extrusion rates than those that contain the cartilage piece. This is the exact opposite of what one would expect, since the sole purpose of the cartilage is to increase fixation and lower extrusion rates. This finding can be attributed in part to the fact that the two data points representing implants without cartilage were taken from the same single study, while the T with cartilage included data from this study as well as another. Had the second study been eliminated, the extrusion rate of the T with cartilage would have been lower than the T without cartilage at 5 years.

Figure 13.4 shows the extrusion rates of the PE systems and bioceramic systems considered in the study. Extrusion rates of all the ceramic systems (solid lines) for which more than five year data were found were consistently better than extrusion rates of all the polyethylene systems studied (dashed lines). Closer examination shows that this is especially so for the bioactive materials (HA, Ceravital®, and Bioglass®). Statistical analysis of two broad groups,

1. pool of all PE prostheses, and
2. pool of all bioactive ceramic prostheses,

shows an extremely significant difference in extrusion rates in favor of bioactive ceramics. This finding is further strengthened by the fact that the bioactive ceramic data in this comparison included extrusions past the five-year mark, as shown in Fig. 13.4.

As stated earlier, one of the primary causes of extrusion is pressure from the implant being exerted on the tympanic membrane, leading to perforation and extrusion, which is controlled primarily by the surgeon during shaping of the part. Based on this factor alone, one would not expect significant differences between different materials (assuming similar surgical technique). However, an additional factor leading to extrusion is fixation of the implant to the surrounding tissue, particularly the mastoid bone if the implant rests there. Unlike pressure exertion of the implant on the tympanic membrane, this factor can be heavily influenced by the type of material used (i.e., bioinert vs. bioactive). This probably accounts for the bioactive materials having much lower extrusion rates than inert materials. Presumably the inherent property of these materials to form a direct chemical bond with the contacting tissue stabilizes the implant and minimizes the incidence of extrusion and/or displacement.

A-B gap

Figure 13.5 shows the A-B gap data. The y-axis gives the percentage of patients with an A-B gap less than or equal to the value given by the abscissa. It was assumed that the data followed a linear relationship. Based

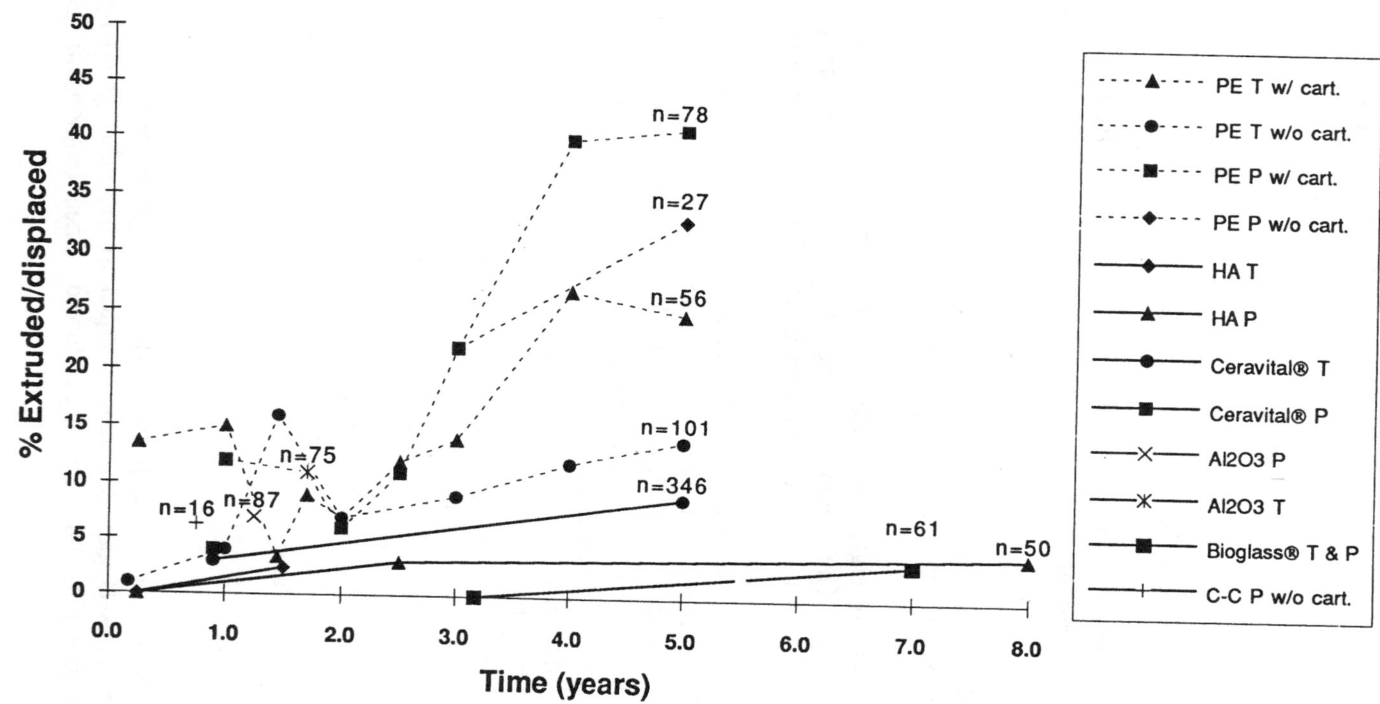

Fig. 13.4 Extrusion/displacement rate vs. time of implantation for various MEP systems.

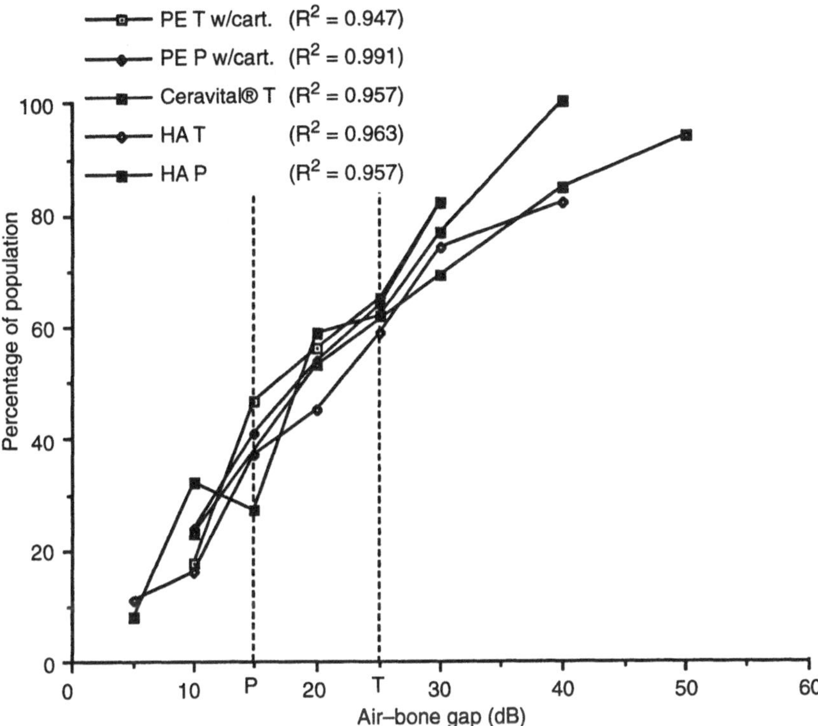

Fig. 13.5 Population distributions of air-bone gaps.

on this assumption and requiring a squared correlation coefficient of greater than 0.9, 5 of the 12 systems studied were considered linear. From the curve-fit equations (see Fig. 13.5) the success rates for the pre-established acceptable A-B gaps (≤ 25 dB for Ts, ≤ 15 dB for Ps) were calculated, from which various statistical comparisons were done.

Table 13.3 shows these comparisons. Significant differences were found between

1. the PE Ts and Ps with cartilage,
2. the Ceravital® Ts and PE Ts with cartilage, and
3. combined HA/PE Ps and HA/PE Ts.

Table 13.3 System comparisons of air-bone gap data to identify significant differences (via Fischer's exact test)

	Success (%)	*Failure (%)*	*Statistical Significance*
PE Ts w/cart.	67.8	32.2	
PE Ps w/cart.	38.6	61.4	*P < 0.0001
HA Ts	55.6	44.4	
HA Ps	38.5	61.5	P = 0.2749
PE Ts w/ cart.	67.8	32.2	
HA Ts	55.6	44.4	P = 0.2643
PE Ps w/ cart.	38.6	61.4	
HA Ps	38.5	61.5	P = 1.0
Ceravital® Ts	56.1	43.9	
PE Ts w/ cart.	67.8	32.2	*P = 0.0267
Ceravital® Ts	56.1	43.9	
HA Ts	55.6	44.4	P = 1.0
All HA	47.2	52.8	
All PE	55.7	44.3	P = 0.2818
HA, PE Ps	40.3	59.7	
HA, PE Ts	65.5	34.5	*P = 0.0009
All Ceravital®	56.1	43.9	
All HA	47.2	52.8	P = 0.2431
All Ceravital®	56.1	43.9	
All PE w/cart.	55.7	44.3	P = 0.9318

Note that unlike extrusion data, there is not an overwhelming advantage in A-B gap closure when using bioactive systems over PE systems, with the exception of the second significant comparison mentioned above, e.g. - Ceravital® Ts and PE Ts with cartilage. Also note that both the first (PE Ts and Ps with cartilage) and third (combined HA/PE Ps and HA/PE Ts) comparisons differentiate the P and T systems. Others have found similar results in favor of T systems (Hughes, 1987; Sanna *et al.*, 1985; Austin, 1984; Giddings and House, 1992; Nikolaou *et al.*, 1992) while others have found opposite results (Colletti *et al.*, 1987; Reck *et al.*, 1988; Smyth, 1982;

Frootko, 1984; Reck and Helms, 1985). It is difficult to identify the reasons for better results from the Ps found here. Quantitatively, otologists have expected and accepted that Ts usually exhibit greater postoperative A-B gaps as this is a more severe reconstruction. As a result, more lenient terms for success are sometimes adopted. Others claim that a procedure using a T is less prone and less sensitive to error than a procedure using a P and therefore it is easier to obtain A-B gaps equal to or better than A-B gaps from P patients. In fact, many surgeons will opt to use a T even when the patient possesses one or two perfectly good ossicles. Continuity of the 'bridge' between the oval window and the tympanic membrane is much better when a T is used.

When expectations of relative performance are eliminated and equal criteria for success are applied to all systems, it must be concluded that there is little difference in hearing results between the systems considered in Fig. 13.5.

CONCLUSIONS

Clinical success rates of ossicular replacement prostheses have been analyzed with respect to extrusion rates and post-operative A-B gaps. The data show that bioactive ceramic systems have much lower extrusion rates than the currently popular porous polyethylene systems, presumably due to better fixation to surrounding tissue. Analysis of the A-B gap data suggests that total ossicular replacement (T) systems exhibit greater audiometric success than partial (P) systems. However, the different criteria for success used for each system must be noted.

Based on these findings, the commonly reported reasons for not using ceramic systems in ossicular replacement, particularly difficulty in shaping the implant to desired dimensions in the operating room, appear to be unwarranted. From the data presented in Fig. 13.5 one can see that by taking a few extra minutes in the operating room to shape a bioactive implant, the need for revision surgery may be cut in half.

Authors should attempt to establish a standard method of reporting their data to allow direct comparisons between studies. This was by far the limiting factor in the analysis of the available data. Finally, as others have previously noted (Smyth, 1982; Jackson, 1983; Smyth, 1984), the need for more long-term clinical data in this field cannot be overemphasized.

REFERENCES

Austin, D.F. (1984) Columellar tympanoplasty, in *Biomaterials in Otology*, (ed J. Grote) Martinus Nijhoff Publishers, Boston.

Babighian, G. (1984) Our experience with Ceravital® implants in middle ear surgery (a middle term evaluation), in *Biomaterials in Otology* (ed J.Grote), Martinus Nijhoff Publishers, Boston.

Blayney, A.W. *et al.* (1986) Ceravital® in ossiculoplasty: experimental studies and early clinical results, *The Journal of Laryngology and Otology*, **100**, 1359-66.

Blayney, A.W. *et al.* (1992) Problems in alloplastic middle ear reconstruction, *Acta Otolarygol,* **112**, 322-27.

Blitterswijk, C.A. *et al.* (1990) The biocompatibility of hydroxyapatite ceramic: a study of retrieved human middle ear implants, *Journal of Biomedical Materials Research*, **24**, 433-53.

Brackmann, D.E. *et al.* (1984) TORPs and PORPs in tympanoplasty: A review of 1042 operations, *Otolaryngology - Head and Neck Surgery*, **92** (1), 32-7.

Brackmann, D.E. and Sheehy, J.L. (1979) Tympanoplasty: TORPS and PORPS, *The Laryngoscope,* **89**, 108-14.

Colletti, V. *et al.* (1987) Minisculptured ossicle grafts versus implants: long-term results, *Am. J. Otology*, **8** (6), 553-59.

Emmett, J.R., Shea, J.J. and Moretz, W.H. (1986) Long-term experience with biocompatible ossicular implants. *Otolaryngology - Head and Neck Surgery*, **94** (5), 611-16.

Epstein, G.H. and Sataloff, R.T. (1986) Biologic and nonbiologic materials in otologic surgery, *Otolaryngologic Clinics of North America*, **19** (1), 45-52.

Frootko, N.J. (1984) Causes of ossiculoplasty failure using porous polyethylene (Plastipore™) prostheses, in *Biomaterials in Otology* (ed J. Grote), Martinus Nijhoff Publishers, Boston.

Gamoletti, R. *et al.* (1984) Histology of extruded Plasti-Pore ossicular prostheses, *Otolaryngology - Head and Neck Surgery*, **92**, 342-45.

Gersdorff, M. *et al.* (1989) Bone allografts in reconstructive middle ear surgery, *Arch Otorhinolaryngol*, **246**, 94-6.

Giddings, N.A. and House, J.W. (1992) Tymplanosclerosis of the stapes: hearing for various surgical treatments, *Otolaryngology - Head and Neck Surgery*, **107**, 644-50.

Goldenberg, R.A. (1992) Hydroxylapatite ossicular replacement prostheses: a four-year experience, *Otolaryngology - Head and Neck Surgery*, **106** (3), 261-9.

Goldenberg, R.A. (1990) Hydroxylapatite ossicular replacement prostheses: Preliminary results, *Laryngoscope*, **100**, 693-700.

Grote, J.J. (1986) Reconstruction of the ossicular chain with hydroxylapatite implants, *Ann Otol Rhinol Laryngol*, **123** (Suppl), 10-12.

Grote, J.J. (1990) Reconstruction of the middle ear with hydroxylapatite implants: long-term results. *Ann Otol Rhinol Laryngol*, **99** (Suppl), 12-16.

Grote, J.J. (1987) Reconstruction of the ossicular chain with hydroxyapatite Prostheses. *Am. J. Otology*, **8** (5), 396-401.

Hench, L.L. and Wilson J. eds. (1993) *An Introduction to Bioceramics*, World Scientific Pub., Singapore.

Hughes, G.B. (1987) Ossicular reconstruction: A comparison of reported results, *The American Journal of Otology*, **8** (5), 371-74.

Jackson, C.G. *et al.* (1983) Ossicular chain reconstruction: the TORP and PORP in chronic ear disease, *Laryngoscope*, **93**, 981-88.

Jahnke, K. *et al.* (1983) Experiences with Al_2O_3-ceramic middle ear implants, *Biomaterials*, **4**, 137-38.

Mangham, C.A. and Lindeman, R.C. (1990) Ceravital® versus Plastipore® in tympanoplasty: A randomized prospective trial, *Laryngoscope*, **99**, 112-16.

Mercandino, E.C. and Tarasido, J.C. (1975) Artificial stapes - a fourteen-year Report, *ORL*, **37**, 169-172.

Merwin, G.E. (1986) Bioglass® middle ear prosthesis: Preliminary report. *Ann Otol Rhinol Laryngol*, **95**, 78-82.

Merwin, G.E., Wilson, J. and Hench L.L. (1984) Current status of the development of Bioglass® ossicular replacement implants, in *Biomaterials in Otology* (ed J. Grote), Martinus Nijhoff Publishers, Boston.

Merwin, G.E., Atkins, J.S., Wilson, J., Hench, L.L. (1982) Comparison of ossicular replacement materials in a mouse ear model. *Otolaryngol Head Neck Surg.*, **90**, 461-69.

Nikolaou, A. *et al.* (1992) Ossiculoplasty with the use of autografts and synthetic prosthetic materials: a comparison of results in 165 cases. *The Journal of Laryngology and Otology*, **106**, 692-94.

Niparko, J.K. *et al.* (1988) Bioactive glass-ceramic in ossicular reconstruction: a preliminary report. *Laryngoscope*, **98**, 822-25.

Palva, T. and Makinen, J. (1983) Histopathological observations on polyethylene-type materials in chronic ear surgery. *Acta Otolaryngol*, **95**, 139-46.

Podoshin, L, Gradis, M. and Gertner, R. (1988) Carbon-carbon middle ear prosthesis: a preliminary clinical human trial report. *Otolaryngology - Head and Neck Surgery*, **99** (3), 278-81.

Portmann, M. *et al.* (1984) Comparative study of different ossicular prostheses in tympanoplasty (Proplast, Plastipore™, Ceravital®). Analysis of clinical results, histopathological and hearing in the long term (250 cases) in *Biomaterials in Otology* (ed J. Grote), Martinus Nijhoff Publishers, Boston.

Reck, R., Storkel, S. and Mayer, A. (1988) Bioactive glass-ceramics in middle Ear surgery. *Ann N.Y. Acad. Sci.*, **523**, 100-106.

Reck, R. (1984) Bioactive glass-ceramics in ear surgery: Animal studies and Clinical results. *Laryngoscope*, **94** (2), 2 Suppl 33, 1-54.

Reck, R. and Helms, J. (1984) Fundamental aspects of Bioglass and surgery with bioactive glass-ceramic implants, in *Biomaterials in Otology* (ed J. Grote). Martinus Nijhoff Publishers, Boston. [***Ed's note: The title of this paper uses the term Bioglass® incorrectly as a generic for bioactive glass. The material actually studied was Ceravital.***]

Reck, R. and Helms, J. (1985) The bioactive glass-ceramic Ceravital in ear Surgery: five years' experience. *The American Journal of Otology*, **6** (3), 280-83.

Sanna, M. *et al.* (1985) Autologous fitted incus versus Plastipore™ PORP in ossicular chain reconstruction. *The Journal of Laryngology and Otology*, **99**, 137-41.

Sanna, M. *et al.* (1984a) Plastipore™ prostheses for ossicular chain reconstruction in tympanoplasty, in *Biomaterials in Otology* (ed J. Grote), Martinus Nijhoff Publishers, Boston.

Sanna, M. *et al.* (1984b) Failures with Plasti-Pore ossicular replacement prostheses. *Otolaryngology - Head and Neck Surgery*, **92** (3), 339-341.

Shea, J.J. (1958) Fenestration of the Oval Window. *Ann. Otol. Rhinol. Laryngol.*, **67**, 932-51.

Shea, J.J. and Emmett, J.R. (1984) Polyethylene TORPS and PORPS in otologic surgery, in *Biomaterials in Otology* (ed J. Grote), Martinus Nijhoff Publishers, Boston.

Shea, J.J. and Homsy, C.A. (1974) The use of Proplast™ in otologic surgery. *Laryngoscope*, **84** (10), 1835-45.

Silverstein, H., McDaniel, A.B., and Lichtenstein, R. (1986) A comparison of PORP, TORP, and incus homograft for ossicular reconstruction in chronic ear surgery. *Laryngoscope*, **96**, 159-65.

Smyth, G.D.L. (1982) Five-year report on partial ossicular replacement prostheses and total ossicular replacement prostheses. *Otolaryngol Head Neck Surg*, **90**, 343-46.

Smyth, G.D.L. (1984) PORPs and TORPs versus allografts after five years, in *Biomaterials in Otology* (ed J. Grote), Martinus Nijhoff Publishers, Boston.

Toner, J.G., Smyth, G.D.L., and Kerr, A.G. (1991) Realities in ossiculoplasty. *The Journal of Laryngology and Otology*, **105**, 529-33.

Wilson, J., Hench, L.L., Greenspan, D.G., eds (1995) *Bioceramics 8,* Elsevier Press, Oxford, England.

Wilson, J., Merwin, G.E. and Hench, L.L. (1985) Machining in Bioglass®. *SAMPE Journal* **21** (3), 6-8.

Yamamoto, E., and Iwanaga, M. (1986) Ossiculoplasty failure with ceramic Ossicular replacement prosthesis. *ORL*, **48**, 332-37.

14

Longevity of Osseointegrated Dental Implants†

Charles F. De Freest
Daniel A. Savett

The use of dental implants as a rehabilitative treatment for edentulism has increased dramatically in the last decade and it is estimated that there are at least 300,000 dental implants placed annually in the United States (NIH, 1988). Dental implants rank third in the global market of implantable biomaterials with an estimated value of $425 million (Kohn, 1992). All of this underscores the need to better review and understand the current status of dental implants.

Endosseous dental implant systems are the focus of this review, however there are other dental implant systems that are currently being used. According to Worthington (1988), endosseous dental implants accounted for a $23.8 million share of the United States' dental implant market, estimated to total $30.8 million in 1987.

Despite the endosseous implant's dominance in the dental marketplace two other major dental implant systems are in use, subperiosteal and transosteal dental implants. The subperiosteal implants take the form of a framework resting on the surface of the jaw bone. They are placed bilaterally in the mandible and support an overdenture. In contrast to the other dental implant systems, there are no claims of subperiosteal implant osseointegration. The frameworks are commonly made of the cobalt-chromium alloy, Vitallium®. Bodine and Yanase (1985) conducted a follow-up of 28 patients with a five year success rate of 90%, a ten year success rate of about 60%, and a 15-year success rate of about 50%. Recently, the same authors updated their study to report a ten year success rate of 79%

®Howmedica Inc, Rutherford, NJ

†The opinions or conclusions contained in this chapter are those of the authors and are not to be construed as official or reflecting the views of the United States Department of Defense or the United States Air Force.

and a 15 year success rate of 60% (Yanase, *et al.,* 1994), however, they concluded that survival rates of subperiosteal implants continue to decrease in the long-term without reaching a steady state. Homoly (1990) has reported on 11-years' experience with subperiosteal implants with a 92% success rate but the details of the criteria for success were not well documented. Many authors report an initially high success with a marked drop after the first five years (Albrektsson and Sennerby, 1991).

Transosteal, or staple, implants are used exclusively in the mandible. They require an external submental incision and are usually placed in an oral surgical suite. Their use has remained fairly constant, at least since 1985 (Worthington, 1988), but accounted for a decreasing proportion of the U. S. implant market. These implants have been fairly successful. Studies by Small and Miziek (1986) report a 90% success rate at ten years, although some mobility and bone loss was noted in 10% of these cases. Other, shorter term studies have been conducted with fairly positive results (Albrektsson and Sennerby, 1991). These implants are generally placed under general anesthesia and a strict surgical protocol which usually precludes their placement in the general dentist's office.

Blade implants are endosseous implants that do not osseointegrate. Blade implants are usually used in areas where the remaining shallow alveolar bone does not allow the use of root-form endosseous implants. The Harvard Blade Implant Clinical Trial (Schnitman *et al.*, 1988), did conclude that blades could be useful but the findings only covered a three year study. Weiss (1988) observed a 91% mandibular blade implant success rate over five to nine years. However, when early failures were included the success rate decreased to 76 %. Although there has been a slow rise in the number of blade implants used in the United States, Worthington (1988) predicted that this would only represent 24% of the total dental implants in use in 1990, being greatly outpaced by the endosseous implants. This seems to be the case.

As stated earlier, osseointegrated root-form endosseous implants represent the greatest share of dental implants placed in the United States and they are the main focus of this chapter. The use of endosseous implants has been recorded in ancient Egyptian and Central and South American civilizations (Balkin, 1988). In the 1800s Maggiolo had fabricated gold endosseous implants (Maggiolo, 1809). Development continued and in 1937, Strock inserted the first cobalt-chromium-molybdenum screw-shaped implant (Strock, 1939). Implant designs and placement techniques advanced throughout the world's private practices with little scientific support by academia (Balkin, 1988). It wasn't until Dr. Per-Ingvar Brånemark introduced the concept of osseointegration and the use of the biocompatible metal, commercially pure titanium, that truly advanced and scientifically supported design and treatment concepts were brought to dental implantology (Brånemark *et al.*, 1985). Throughout the remainder of this chapter the term endosseous implant will refer only to those implants which are expected, due to their design and materials fabrication, to become osseointegrated.

There are four major endosseous systems available: titanium, titanium alloy, coated titanium alloy, and all-ceramic. Brånemark (Nobelpharma USA, Inc.), Core-Vent® (Core-Vent, Inc.), Interpore IMZ® (Interpore, Intl.), Integral® (Calcitek, Inc), Steri-oss® (Bausch & Lomb, Inc), and Stryker Precision® (Stryker, Inc) are six popular cylindrical endosseous dental implants systems that represent one or more of these endosseous implant categories and have their own reported advantages and disadvantages (Christensen, 1990). However, this is by no means a complete list and the dentist can easily become overwhelmed if not well educated in this evolving field.

Because of its strength and biocompatibility, titanium has become the material of choice in endosseous dental implants. Titanium alloys of aluminum and vanadium (Ti-6Al-4V) are also used. Both commercially pure titanium and Ti-6Al-4V have excellent corrosion resistance under physiological conditions (Kohn, 1992). Although relatively inert there are reports of ion release as a result of chemical dissolution of the oxide layer (Healy and Ducheyne, 1992). It is still unclear what, if any, are the consequences of this dissolution on implant osseointegration and the health of the patient. Efforts have been made to enhance osseointegration prior to loading and, as a consequence, various coating methods have been developed for the titanium implants. Calcium phosphate coatings, such as hydroxyapatite, have been very popular and studied extensively. Because of the large range of compositions available Ducheyne (1987) proposed the term 'calcium phosphate ceramics' instead. There are very important physical properties of these ceramics that are often left out of sales literature such as, (but not limited to) pore size, pore shape, phases present, percentage crystallinity and coating thickness. The discrepancies in clinical data on these coated implants suggest that material and processing induced changes in both the ceramic and the metal affect implant performance (Kohn, 1992).

Single-crystal sapphire implants are one-stage aluminum oxide ceramic implants, containing no metal copings. These endosseous implants have not been used to any great extent in the United States by practicing dentists but have been used in Japan and Sweden with some reported successes (Albrektsson and Sennerby, 1991).

NIH CONSENSUS CRITERIA (from: NIH Consensus Development Conference: June 1988)

In 1987, for the years 1983 through 1987, the National Institutes of Health (NIH) estimated that the number of dental implants placed in the United States had grown fourfold. At the same time the number of practitioners placing dental implants had grown tenfold. Furthermore, it was estimated that by 1992 the number of dental implants placed in the U.S. would grow to three hundred thousand. The major reason for this growth in numbers of implants placed was the ever increasing popularity of the endosseous dental implant as a means to treat edentulism, especially for those patients who had difficulties in wearing traditional dental prostheses.

Despite an overall decline in decayed, missing, and filled teeth which had been observed over several decades, edentulism in adults beyond thirty-five years of age remained, and remains today, a significant problem (Brown, 1994). While many partially and fully edentulous individuals can wear traditional dental prostheses, some have extensive loss of tooth-bearing or prosthesis-bearing bone or are medically compromised to the point where they are unable to manage conventional prostheses. It is for these patients that dental implants offer their greatest benefit.

In June, 1988 NIH convened a consensus development conference in which an attempt was made to assess, from studies available up to that date, the overall safety and efficacy of dental implants. A representative group consisting of clinicians, researchers, and educators, most of whom were involved in recent developments in dental implantology, formulated a consensus on five separate questions.

The first was 'What is the evidence that dental implants are effective for the long term?' A summary of their opinions on this particular question is:

- The criteria for success vary for different implant systems.

- It is quite difficult directly to compare different types of implants since indications and success criteria differ.
- Poor research designs prohibited proper comparison of studies involving comparable implants. Randomized, controlled trials had not been used.
- The panel could only conclude that the evidence from a number of case studies provided sufficient evidence to support the conclusion that 'when specific types of implants are inserted by clinicians experienced with the respective techniques, a large proportion of implants remain in place for periods of ten years or more.'
- Definite statements regarding long-term success could not be made, due mainly to unreported information and lack of uniform application of proper research designs in the studies that were evaluated.

The group developed a set of recommendations for the conduct of future research in order to resolve the problems observed with the existing studies. *'Future case studies should conform to the following principles:*

- *A prospective statement of study aims, with clear definitions of success and failure for all measures.*
- *A description of the study populations and criteria for patient selection.*
- *Standardization, to the extent possible, of treatment outcome measures, with presentation of data on reliability. Use of independent examiners is desirable.*
- *Adequate sample size adjusted for the expected attrition over the length of the study.*
- *Concise reporting of reasons for attrition.*
- *Reporting of all failures from time of insertion of the implants.*
- *Documentation and follow-up of each failure.*
- *Use of standardized reporting methods, including life tables.*
- *Limiting extrapolation of results to populations similar to that of the study under similar experimental conditions.'*

Finally, it was noted that important information on traditional treatment outcome measures such as patient satisfaction, comfort, masticatory function, esthetics, phonetics, and absence of both physical and psychological symptoms was frequently not presented.

OUR METHOD OF REPORTING IMPLANT STUDY RESULTS

Our review of numerous studies revealed that very few satisfied the proposed guidelines set forth by the NIH Consensus Development Conference. A summary of each study is presented as a table. Each column represents a separate category of information as follows:

COLUMN A: Study reference number.

COLUMN B: Year the study was reported.

COLUMN C: Study length, the total period of observation for all implants evaluated within the study.

COLUMN D: Type of implants studied. The following abbreviations are used.

Br	Brånemark, Titanium screw
IMZ	IMZ, Titanium cylinder
ITI	ITI, Hollow Titanium Basket
CV	Core Vent, TiAl6V4 cylinder with coronal threads, Screw Vent/Swede Vent
WD	Wide Diameter, Titanium screw
IN	Calcitek Integral, HA coated cylinder
SO	Steri-Oss, HA coated screw & cylinder

COLUMN E: Number of implants placed initially in study, i.e., total placed.

COLUMN F: Number of implants followed through to the end of the study.

COLUMN G: Percentage of total number of implants placed which were lost during the study period because follow-up was not done.

COLUMN H: Number of patients initially entered into the study.

COLUMN I: Number of patients actually followed through to the end of the study.

COLUMN J: Percentage of total number of patients entered into the study which were lost during the study period because follow-up was not done.

COLUMN K: Number of males initially entered into the study.

COLUMN L: Males' age range.

COLUMN M: Males' average age.

COLUMN N: Number of females initially entered into the study.

COLUMN O: Females' age range.

COLUMN P: Females' average age.

COLUMN Q: Number of implants placed in both the anterior region of the maxilla and the anterior region of the mandible. This includes implants placed from the anterior wall of one maxillary sinus to the anterior wall of the other maxillary sinus as well as implants placed between the mandibular mental foramina. Where a study did not distinguish between anteriorly placed and posteriorly placed implants, the distinction was usually made between placement in either the maxillary or mandibular arch. The symbol < > will denote data for the maxillary arch as a whole. Where information is given for both the anterior maxilla and the whole maxilla, two numbers will appear.

COLUMN R: Number of implants failed in anterior maxilla and anterior mandible and/or < Number of implants failed in maxillary arch>; includes implants not accounted for at study's end for any reason.

COLUMN S: Percentage of implants succeeding in anterior maxilla and mandible and/ or <Percentage of implants succeeding in maxillary arch>.

COLUMN T: Number of implants placed in both the posterior region of the maxilla and the posterior region of the mandible. This includes implants placed distal to the anterior walls of the maxillary sinuses as well as implants placed distal to the mandibular mental foramina. Where a study did not distinguish between anteriorly placed and posteriorly placed implants, the distinction was usually made between placement in either the maxillary or mandibular arch. The symbol < > will denote data for the mandibular arch as a whole. Where information is given for both the posterior mandible and the whole mandible, two numbers will appear.

COLUMN U: Number of implants failed in posterior maxilla and posterior mandible and/ or < Number of implants failed in mandibular arch>; includes implants not accounted for at study's end for any reason.

COLUMN V: Percentage of implants succeeding in posterior maxilla and mandible and/ or <Percentage of implants succeeding in mandibular arch>

COLUMN W: Total success recorded as an average success between anterior and posterior areas of the dental arches or < as an average percent of success between both the maxilla and mandible>. In either case, this is a total mouth average disregarding any one particular area and is given only because studies presented results in this fashion.

COLUMN X: Restorations placed on implants

F	Fixed Prosthesis
P	Removable Partial Denture Prosthesis
C	Removable Complete Denture Prosthesis

COLUMN Y: NIH Proposed Criteria followed in the Study. Letter(s) in this column will indicate which criteria were adhered to by the authors. We applied the rules leniently.

A	A prospective statement of study aims, with clear definitions of success and failure for all measures.
B	A description of the study populations and criteria for patient selection.
C	Standardization, as far as possible, of treatment outcome, with presentation of data on reliability. Use of independent examiners is desirable.
D	Adequate sample size adjusted for the expected attrition over the length of the study.
E	Concise reporting of reasons for attrition.
F	Reporting of all failures from time of insertion of the implants.
G	Documentation and follow-up of each failure.
H	Use of standardized reporting methods, including life tables.
I	Limiting extrapolation of results to populations similar to that of the study under similar experimental conditions.

DISCUSSION OF FINDINGS FROM DENTAL IMPLANT LONGEVITY STUDIES

A total of twenty-five current journal articles was evaluated to provide the data in Table 14.1. These are listed as <u>chart references</u> at the end of this chapter and are designated by superscript numerals. We found that most

Table 14.1 Dental implant longevity studies

Reference Number	Year Study Published	Study Length (Years)	Implant Type	# Implants Placed Initially	# Implants Followed to Completion	% Implants Lost to Follow-up	# Patients Followed Initially	# Patients Followed to Completion	% Patients Lost to Follow-up	# Males Followed	Male Age Range	Male Age Average
1	1993	1 to 3	Br	558	460	18	159	139	13	67	18 to 70	
2	1994	2 to 6	Br, IMZ	190	190	0	67	67	0	41		
3	1994	1 to 3.6	Br	137	137	0	137	137	0	34		
			IMZ	497	497	0	497	497	0	123		
			CV	115	115	0	115	115	0	29		
4	1994	1 to 2	WD	154	154	0	50	50	0			
5	1993	2 to 8	IN	1374			427			162		
6	1993	1 to 5	IN	745	9	97						
7	1993	1 to 8	IN	690	690	0	221	221	0	112	9 to 86	53.6
8	1993	5	Br, ITI	142	142	0	67	59	22	17		68.9
9	1993	2 to 8	Br	40	40	0	32	32	0	17	16 to 63	30.8
10	1993	2 to 8	Br	94	94	0	30	30	0	15	19 to 56	40.8
11	1993	3 to 8	Br	105	105	0	35	35	0	11	31 to 64	47.0
12	1994	1 to 2	Br	72	72	0	36	36	0	17	36 to 85	63.7
13	1993	0 to 7	Br	166	634*	7*	215*	198*	8*	69		56.0
			IMZ	517	634*	7*	215*	198*	8*			
14	1993	1 to 4	Br	254	238	6	48	39	19	18	80 to 87	82.7
15	1992	0.6 to 8	Br	673	673	0	169	169	0	36	21 to 83	
16	1992	0.5 to 2	CV	85	85	0	31	31	0			
			Br	107	107	0	25	25	0	8	19 to 72	53.0
17	1993	1 to 5	IMZ	513	491	4.3						
18	1992	1 to 5	SO	670	462	31	280	280	0	112	16 to 84	
19	1990	4 to 9	Br	274	274	0	46	46	0	10	35 to 63	49.9
20	1990	10 to 15	Br	4636			700	557	26	302	19 to 79	55.3
21	1989	5	CV	1732	1605	7	623	600	4	232	16 to 81	
22	1988	3	IN	815	815	0	367	367	0			
23	1987	0 to 4.3	Br	410	410	0	133	133	0		24 to 81	
24	1986	2	Br	358	358	0	70	70	0	20	20 to 78	55.0
25	1981	1 to 15	Br	2768	1997	28	371	284	23	38%		

Both Br and IMZ were combined in this study.

# Females Followed	Female Age Range	Female Age Average	# Implants Anterior <Maxillary>	# Failed Anterior <Maxillary>	% Success Anterior <Maxillary>	# Implants Posterior <Mandibular>	# Failed Posterior <Mandibular>	% Success Posterior <Mandibular>	Total % Success Anterior+ Posterior <Max. + Mand.>	Type of Prosthesis	NIH Criteria Followed
92	18 to 70		<220>	<16>	<93>	<338>	<17>	<95>		F	E, F, H
26			<82>			<108>			<85>	C	
103									<96>	C	
374									<96>	C	
86									<88>	C	
									<95>	F	
265					95 <94>			94 <95>	93		
			<393>	<9>	<98>	<352>	<5>	<99>	<98>		H
109	9 to 86	53.6	<327>	<13>	<96>	<363>	<11>	<97>	<97>		E, F
42		62.2							<91 & 92>	C	E, F
15	17 to 64	35.7	27 <28>	0 <0>	100 <100>	12 <12>	0<0>	100 <100>	100 <100>	F	A–C, E, G–I
15	19 to 62	41	94<50>	8 <3>	91 <94>	<44>	<5>	<89>	<91>	F, C	A–C, E, G–I
24	20 to 65	45	<41>	<1>	<98>	105 <64>	6 <5>	94 <92>	<94>	F	A–C, E, G–I
19	36 to 85	63.7	72	0	100	<72>	<0>	<100>	100	C	A, C, E, I
146		56	<69>						<87>	F, C	H
			<115>							F, C	H
30	80 to 89	82.7	<57>	<3>	<95>	<197>	<0>	<100>	<97>	F, C,	A, B, E, G, I
133	21 to 83		572	24	96	101	7	93		F, C	
			<62>	<9>	<85>	<23>	<0>	<100>		F	A, B, C, E–G
17	19 to 72	53	<56>	<3>	<95>	<51>	<0>	<100>			
				<17>			<5>		<97>		A, C, E
168	15 to 78		<205>	<20>	<90>	<257>	<14>	<95>	<93>	F,C	A–D, F, H, I
36	28 to 69	49.8		<36>	<1>	<97>	<238>	<38>	<84>		A, B, C, E–G
398	19 to 79	55.3	<1789>			<2847>			<84 to 92>	F, C	C, H
391	16 to 81		<601>	<23>	<96>	<923>	<29>	<97>		F, C	
			<426>			<389>			<97>		
	24 to 81		410 <189>	20 <14>	93	<221>	<6>	<97>		F	A, B, E, F
50	20 to 78	55	<17>	<0>	<100>	<269>	<3>	<99>		F	B, E, F, G
62%			<981>	<244>	<75>	<1016>	<120>	<88>	<82>	F, C	A, C–H

studies done before 1985, when endosseous implants came into widespread use within the dental profession, were continued into this decade and have been updated, with data from recall of the same patients and others added to the study populations. Because almost all studies we found were designed to present information from consecutively treated patients within a clinical practice, the study lengths appear as a range of years rather than a single year. Ultimately, this method of reporting implant survival statistics on a continuum basis involves the use of time-dependent life table analyses (Lill, 1993). The difficulty in interpreting data presented in this way is that,

- Different covariates are used for different studies.
- Many authors do not explain the loss of subjects (and their implants) during the study period.
- Many studies attempt to use these tables to summarize very complex sets of interacting data, most of which is confounded by uncontrolled variables, and may lead the reader to believe that the survival statistics presented for dental implants are high for most areas of the oral cavity.

We believe that the method of reporting survival statistics must be standardized so that potential abuse is limited and statistics may be compared with confidence.

Seven different implant systems were evaluated in the studies we have cited. The Brånemark, commercially pure titanium, screw system is by far the most studied. The only other commercially pure titanium screw implants that were evaluated were the Steri-Oss[18] and the Wide-diameter.[4] Only one reference for each of these systems was found, however, the Wide-diameter has been very recently introduced to market. Its supposed, although unproven advantage, is to provide a greater surface area of bone contact where a short fixture is needed. One study[8] compared Brånemark fixtures with the ITI system, which has a commercially pure titanium hollow basket design. The Core-Vent system[3,16,21] consists of Ti-6Al-4V alloy screws and hollow cylinders with partial threads. One study[3] compared the Core-Vent with the IMZ and Brånemark systems. Although its success was much lower, (88% vs. 96% for both IMZ and Brånemark systems,) the authors admitted that they had placed a much higher percentage of the Core-Vent implants into the generally lower quality bone of the posterior maxilla. Also, we note that only 24 Core-Vent implants were placed in the maxilla, whereas 401 Brånemark and IMZ implants, were placed in the mandible. This is clearly an unfair comparison. The Steri-Oss system consists of the full spectrum of available designs, including titanium and HA-coated titanium in cylinder and screw forms. One study[18] compared the various Steri-Oss implant designs with one another. The Calcitek Integral HA-coated cylinder implant system is the most studied HA-coated system and has been on the market the longest.[5,6,7,22]

There seem to be no appreciable differences in success rates among the various implant systems.

In many of the references, there is either a reported or assumed zero percentage loss of patients to follow-up throughout the studies. The assumption is based upon a reported number of implants placed and reported upon, with no data on drop-out. We feel that it is unlikely that all the initially placed implants were carried over to the end of the study. However, there were some interesting contrasts. A Loma Linda study[6] had 97% of the initially placed implants lost to follow-up by the fifth year of the study yet their life tables report 98-99% success rates. One may easily and rightly question these rates when, after five years, only nine implants are left in the study from the original 745 implants. Brånemark's fifteen year study[25] reported a 28% loss to follow-up, which is more believable. Although many studies reported an age range for their male and female populations, it is difficult to calculate the age distributions. One study[7] had male and female age ranges from 9 to 86 years with a mean age of 53.6 years while another study[24] reported an age range of 20 to 78 years with a mean of 55 years. Although the means are quite close, the ranges are not, which reflects the ambiguity in these reporting methods. The bone quantity and quality in a 20 year old will be quite different from that of an 86 year old, and have a great influence on successful osseointegration. Most studies fail to document the bone types into which implants are placed. A highly cortical type II bone has a greater potential for success than thin cortical, highly cancellous type IV bone. Fugazzotto[17] evaluated IMZ implants placed only in type IV bone but failed to give adequate population statistics. The most obvious difficulty with the statistical reporting is the potential for gender bias. In most every study reviewed, females outnumbered males by up to 2:1. Despite this, none of the studies reported separate statistics for the two genders. These studies say nothing about the consequences of placing implants in the osteoporotic bone of postmenopausal women (Baxter, 1993).

In sections Q through W we assess implant success statistics based upon implant location. While it is generally accepted that higher quality bone and greater chance of success will be found in the anterior maxilla and mandible rather than the corresponding posterior regions, few studies separated anterior from posterior implant success statistics. Most studies reported statistics for the maxilla versus the mandible. For the maxilla, success ranged from 75% to 100% for all studies. However, for more recent studies, success ranged from 90% to 100%. Success in the mandible was 84% to 100% and did not change in the recent studies. The change in the results in the maxilla recently may reflect the improved ability of surgeons to deal with type IV bone which is predominant in the posterior maxilla. When anterior maxilla and mandible statistics are evaluated together, reported success ranged from 91% to 100%. Success in the posterior maxilla and mandible combined, range from 93% to 100%. Based upon the statistics from these studies, implant therapy cannot yet be recommended to a patient with confidence because of the variability with which the statistics have been compiled.

Multiple restorative options exist for patients provided with implant therapy. Partially edentulous patients will usually receive a single or multiple unit, screw-retained, fixed prosthesis. Completely edentulous patients can be treated with either a multiple unit, screw-retained fixed prosthesis, a screw-retained bar with clip-retained complete overdenture, or a conventional overdenture, where individual implants carry separate attachments. While most studies that discussed prosthesis design stressed the importance of factors such as occlusal scheme and load transfer to individual implant fixtures, none adequately separated statistics sufficiently to where we could determine the relative effect of different prosthesis designs on implant success.

Finally, we attempted to determine whether or not individual studies conformed to the criteria for dental implant studies set forth by the NIH. We applied the criteria quite liberally but found no study conformed completely. While most studies adequately stated their purpose, their specific aims were often poorly supported by the actual study designs. All studies were inadequately controlled with respect to age and gender balance. Criteria for success and failure differed between studies. Prospective study populations often consisted of patients who walked into the clinic and ultimately elected implant therapy as a treatment option. This does not provide an adequate population balance. None of the prospective studies employed independent examiners. In fact, most of the large studies involving the Brånemark implant system have been conducted by individuals who were directly involved in the system's initial development. None of the prospective studies projected a sample size but continued to add subjects to their populations over time with no set limits. Most of the studies did adequately explain loss of patients from the study populations, but it was often difficult to determine the associated number of implants and their locations lost to follow-up. Life tables were frequently used to report success rates over time but the construction methods used by various authors vary and must be standardized before viable comparison between studies can be made. None of the studies inappropriately extrapolated their results, but several concluded that their implant systems were highly successful and viable treatment modalities when there were too many uncontrolled variables to support such conclusions.

REFERENCES

Albrektsson, T. and Sennerby, L. (1991) State of the art in oral implants. *J Clin Periodontol,* **18**, 474-81.

Balkin, B. (1988) Implant dentistry: Historical overview with current perspective. *J Dent Ed,* **52**, 683-85.

Baxter, J. and Fattore, L. (1993) Osteoporosis and osseointegration of implants. *J Prosthodont,* **2**, 120-25.

Bodine, R.L., and Yanase, R.T. (1985) Thirty year report on 28 implant dentures inserted between 1952 and 1959. *Int Symp Preprosth Surg,* Palm Springs, CA.

Brånemark, P., Zarb, G.A. and Albrektsson, T. (1985) Tissue integrated prostheses: osseointegration. *Clinical Dentistry,* Quintessence Publishing Chicago.

Brown, L. (1994) Trends in tooth loss among U.S. employed adults from 1971 to 1985. *JADA,* **125**, 533-40.

Christensen, G. (1990) Implant prosthetics contribute to restorative dentistry. *JADA*, **121**, 340-53.

Ducheyne, P. (1987) Bioceramics: Materials characterization v. *in vivo* behavior. *J Biomed Mat Res,* **21**, 219.

Healy, K.E. and Ducheyne, P. (1992) The mechanisms of passive dissolution of titanium in a model physiological environment. *J Biomed Mater Res,* **26**, 319-38.

Homoly, P.A. (1990) The restorative and surgical technique for the full maxillary subperiosteal implant. *JADA,* **121**, 404.

Kohn, David. (1992) Overview of factors important in implant design. *Journal of Implantology*, **18**, 204-19.

Maggiolo (1809) in *Manual de l'art dentaire*. (ed. Nancy) C. Leseure, Paris.

NIH consensus development conference statement - dental implants. (1988). *J Oral Implantol,* **14** (2), 127-35.

Schnitman, P.A., Rubenstein, J.E., Whorle, P.S., *et al.* (1988) Implants for partial edentulism. *J Dent Ed,* **52**, 725-36.

Small, I. and Miziek, D. (1986) 16-year evaluation of the mandibular staple bone plate. *J Oral Maxillofac Surg,* **44**, 60-6.

Strock, E. (1939) Experimental work on a method for the replacement of missing teeth by direct implantation of a metal support into the alveolus. *Am J Orthodont Oral Surg*, **25**, 467.

Weiss, M.B. and Rostoken, W. (1982) Development of a new endosseous dental implant. *J Prosthet Dent,* **47**, 633-45.

Worthington, P. (1988) Current implant usage. *J Dent Ed*, **52**, 682-95.

Yanase, R. *et al.* (1994) The mandibular subperiosteal implant. *J Prosthet Dent,* **71**, 369-74.

REFERENCES TO TABLE 14.1

1. Henry, P.J., Tolman, D.E. and Bolender, C. (1993) The applicability of osseointegrated implants in the treatment of partially edentulous patients: Three-year results of a prospective multicenter study. *Quintessence Int*, **24**, 123-9.

2. Mensdorff-Pouilly, Haas, R., Mailath, G. and Watzek, G. (1994) The immediate implant: A retrospective study comparing the different types of immediate implantation. *Int J Oral Maxillofac Implants*, **9**, 571-578.

3. Cune, M.S., de Putter, C. and Hoogstraten, J. (1984) Treatment outcome with implant retained overdentures: Part I - Clinical findings and predictability of clinical treatment outcome. *J Prosthet Dent*, **72** (2), 144 51.

4. Siddiqui, A. and Caudill, R. (1994) Proceedings of the Fourth International Symposium on Implant Dentistry: Focus on esthetics. *J Prosthet Dent,* **72** (6), 623-34.

5. Block, M.S. and Kent, J.N. (1993) Cylindrical HA-coated implants - 8-year observations. *Compend Contin Educ Dent*, **15:S,** 525-32.

6. Lozada, J.L. and James, R.A. (1993) HA - Coated implants: Warranted or not? *Compend Contin Educ Dent,* **15:S**, 540-43.

7. Guttenberg, S.A. (1993) Longitudinal report on hydroxyapatite-coated implants and advanced surgical techniques in a private practice. *Compend Contin Educ Dent,* **15:S**, 549-53.

8. Mericske-Stern, R. and Zarb, G.A. (1993) Overdentures: An alternative implant methodology for edentulous patients. *Int J Prosthodont*, **6**, 203-8.

9. Schmitt, A. and Zarb, G.A. (1993) The longitudinal clinical effectiveness of osseointegrated dental implants for single tooth replacement. *Int J Prosthodont*, **6**, 197-202.

10. Zarb, G.A. and Schmitt, A. (1993) The longitudinal clinical effectiveness of osseointegrated dental implants in anterior partially edentulous patients. *Int J Prosthodont,* **6**, 180-88.

11. Zarb, G.A. and Schmitt, A. (1993) The longitudinal clinical effectiveness of osseointegrated dental implants in posterior partially edentulous patients. *Int J Prosthodont*, **6**, 189-96.

12. Naert, I., Quirynen, M. Hooghe, M. and van Steenberghe, D. (1994) A comparative prospective study of splinted and unsplinted Brånemark implants in mandibular overdenture therapy: A preliminary report. *J Prosthet Dent,* **71**, 486-92.

13. Lill, W., Thornton, B., Reichsthaler, J. and Schneider, B. (1993) Statistical analyses on the success potential of osseointegrated implants: A retrospective single-dimension statistical analysis. *J Prosthet Dent*, **69**, 176-85.

14. Jemt, T. (1993) Implant treatment in elderly patients. *Int J Prosthodont*, **6**, 456-61.

15. Drago, C.J. (1992) Rates of osseointegration of dental implants with regard to anatomical location. *J Prosthodont,* **1**, 29-31.

16. De Bruyn, H., Collaert, B., Linden, U., Flygare, L. (1992) A comparative study of the clinical efficacy of Screw Vent implants versus Brånemark fixtures, installed in a periodontal clinic. *Clin Oral Impl Res,* **3**, 32-41.

17. Fugazzotto, P., Wheeler, S. and Lindsay, J. (1993) Success and failure rates of cylindrical implants in type IV bone. *J Periodontol,* **64**, 1085-87.

18. Saadoun, A. LeGall, M. (1992) Clinical results and guidlines on Sterioss endosseous implants. *Int J Periodonal Rest Den,* **12**, 487-99.

19. Zarb, G.A. and Schmitt, A. (1990) The longitudinal clinical effectiveness of osseointegrated dental implants: The Toronto study. Part I: Surgical results. *J Prosthet Dent,* **63**, 451-57.

20. Adell, R. Eriksson, B., Lekholm, U., Brånemark, P-I., Jemt, T. (1990) A long-term follow-up study of osseointegrated implants in the treatment of totally edentulous jaws. *Int J Oral Maxillofac Implants,* **5**, 347-59.

21. Patrick, D., Zosky, J., Lubar, R., Buchs, A. (1989) The longitudinal clinical efficacy of Core-Vent dental implants: A five-year report. *J Oral Implantol,* **15** (2), 95-103.

22. Golec, T.S. (1988) Three year clinical review of HA coated titanium cylinder implants. *J Oral Implantol,* **14** (4), 437-54.

23. Van Streenberghe, D., Quirynen, M., Calberson, L., Demanet, M. (1987) A prospective evaluation of the fate of 697 consecutive intra-oral fixtures modum Brånemark in the rehabilitation of edentulism. *J Head & Neck Pathol,* **6**, 53-58.

24. Laney, W. *et al.*, (1986) Dental implants: Tissue-integrated prosthesis utilizing the osseointegration concept. *Mayo Clin Proc,* **61**, 91-97.

25. Adell, R., Lekholm, U., Rockler, B. and Brånemark, P.I. (1981) A 15-year study of osseointegrated implants in the treatment of the edentulous jaw. *Int J Oral Surg,* **10**, 387-416.

15

Alveolar Ridge Maintenance Implants

H. R. Stanley
A. E. Clark
L. L. Hench

INTRODUCTION

In 1984 it was estimated that there were 20 million totally edentulous persons in the U.S. Of those wearing dentures 70% were dissatisfied with their mandibular dentures (Misch, 1984). Continuous resorption of the residual alveolar ridge after dental extraction was a major contributor to the dissatisfaction. Alveolar ridge bone loss in these patients reduces denture stability and retention and produces impaired masticatory efficiency, oral and systemic health problems, and compromised esthetics (Stanley, 1987; Quinn and Kent, 1984). These complications are especially severe in elderly patients who have been edentulous for many years (Bell, 1986). Dentists agree that many patients will never wear a removable mandibular complete denture with any degree of satisfaction (Rothstein, 1984). Clinically, resorption can progress so rapidly that dentures cannot be worn for more than a short period of time before a reline or rebase is necessary or, in later years, there may be insufficient alveolar bone remaining for any denture retention (Tallgren, 1967, 1969). Extreme alveolar bone resorption has been termed a 'major disease entity' for the elderly (Atwood, 1971).

According to Bell (1986) loss of natural teeth initiates a remodeling of the residual alveolar ridge. Following extraction of teeth, the dental alveoli fill with blood, which subsequently clots and is replaced with new bone. The contour of alveolar bone undergoes continuous change (bone resorption and subsequent structural rearrangement) according to the degree of stress applied. The total amount of bone resorbed and the rate of resorption not only are different for each individual, but also vary greatly in the same individual at different times (Sobolik, 1980; Atwood, 1962; Veldhuis *et al.*,

1984). The reduction of residual alveolar ridges occurs most rapidly in the first 6 to 24 months after extraction, but in many individuals it appears to continue until death, resulting in the removal of massive amounts of bone (Quinn, *et al.*, 1985; Atwood, 1979). Height loss in the first five years is more than twice the height loss in the succeeding 20 years (7.6 mm/3.1 mm) (Veldhuis *et al.*, 1984). However, there is not much difference after 10 years. The resorption rate of the mandible averages four times the rate of the maxillae (Quinn and Kent, 1984). Even when wearing full dentures, the resorption rate of the mandibular ridge is three times greater than the maxilla (Tallgren, 1967, 1969).

Various techniques have been used in an attempt to preserve or rebuild the edentulous alveolar ridge, mostly with limited success (Kwon *et al.*, 1986). Since the loss of tooth roots sometimes cannot be avoided because of the extent of caries and periodontal disease, alloplastic implants have been used as a substitute for natural roots for many years (Stanley *et al.*, 1995). Lam, (1972) suggested that an implant in the socket of an extracted tooth might simulate a tooth root and preserve the alveolar bone. Since loss of natural tooth roots is the basic cause of various complaints, the most straightforward approach to solving this problem is to replace natural tooth roots with artificial implants (Dennissen *et al.*, 1978).

The ideal implant material for maintaining the alveolar bone should meet the following criteria: (1) no evidence of early resorption of the implant material; (2) acceptable strength to fill the space without crushing under masticatory forces; (3) strong attachment to the soft and hard tissues at the implant interface; and (4) no adverse host reactions. Numerous researchers have found that implants, which act as space fillers after extraction of tooth roots, do delay resorption of residual alveolar ridges (Stanley, 1995). Implants provide mechanical support as a scaffolding and prevent the collapse of both the labial and lingual plates of bone (Dennissen and de Groot, 1979; Stanley *et al.*, 1987).

Various synthetic products have been tried; acrylic resin, hydroxyapatite, coralline hydroxyapatite, carbon, calcium phosphate ceramics, tricalcium phosphate, and a bioactive glass (Bioglass®). Of these, dense hydroxyapatite (HA) and Bioglass® cones offer the greatest potential. The relative clinical performance of these two alloplastic materials is discussed in this chapter.

HYDROXYAPATITE CONE IMPLANTS

Based upon their previous work, Denissen and de Groot (1979) placed 50 non-biodegradable dense calcium hydroxyapatite (HA) cones into the empty sockets of mandibular premolars of dogs. The wounds were permitted to heal by primary closure of the mucosa. Radiographic studies showed the surrounding bone to be closely adapted to the implant after three months, even over the implants. All implants were retained up to 18 months.

Also, in 1979-1980, Denissen and de Groot placed 100 dense HA ceramic cone implants into 20 patients with severe periodontal involvement or teeth with draining fistulae following endodontic treatment and apical curettage. The cones were placed in fresh sockets and closed before the seating of either complete dentures, or fixed or removable prostheses. Nine of the 20 patients received immediate lower, mandibular dentures. Implants were inserted in such a way that their cervical plane was situated just below the most apical part of the socket crest (mostly the vestibular buccal part) to permit primary closure of the extraction wounds. After one year all implants were still in place without radiolucencies. The residual alveolar bone closely adhered to the implants even in those patients who received immediate dentures. Seventy-one implants were placed in nine patients under mandibular full dentures. The physical presence of the implants maintained a bulky ridge; the ridges collapsed only at sites where no implants were present.

Sixty nine implants were retained at 18 months, as reported by Denissen *et al.* (1980) (Fig. 15.1). One was removed at 18 months due to a long-term dehiscence (exposure of the implant through the mucosa). A total of six dehiscences occurred under mandibular full dentures up to 24 months. This process could be arrested by shortening the implant. No dehiscences recurred up to 12 months following shortening. Even those implants requiring shortening appeared to be very strongly attached to the bone at the time of surgery. It appeared that the cones became ankylosed as would natural roots. Denissen *et al.*, (1980) continued to follow these patients up to 30 months (Fig. 15.1).

It seemed to Denissen *et al.* (1980) that beyond doubt the physical presence of such implants, simulating the principle of an ankylotic root, prevented collapse of the cortical plates and guaranteed a residual bulk or

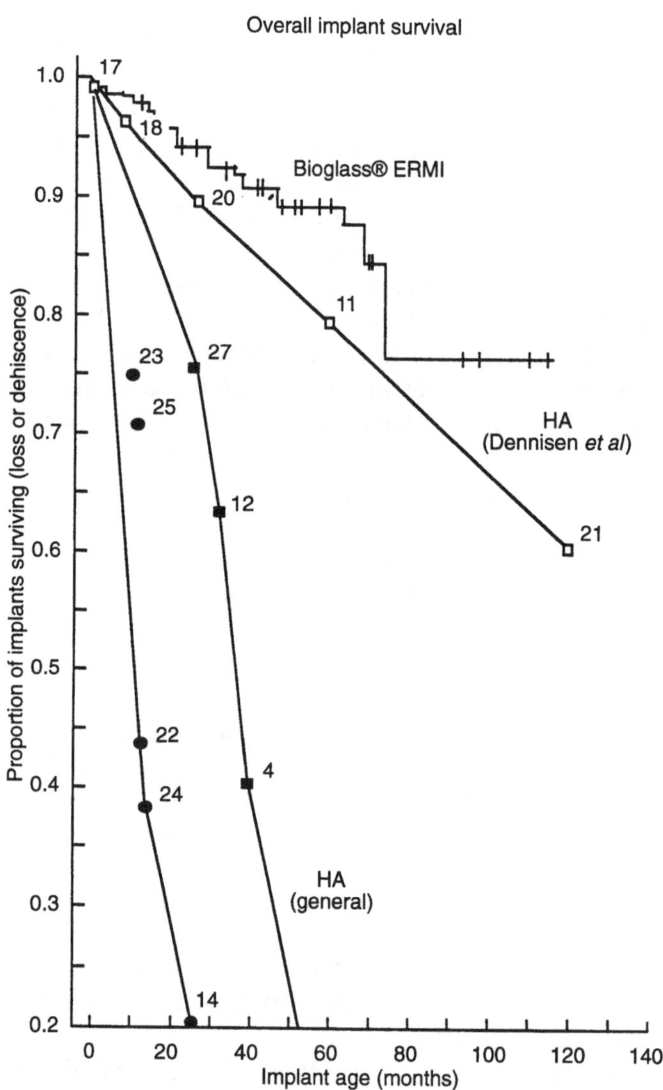

Fig. 15.1 Survivability of alveolar ridge maintenance implants. Numbers correspond to reference citations given in bold.

volume of the denture-bearing region of the mandible, whether the implants were submucosal or permucosal. They recommended that the implants be tightly wedged at least 2 mm below the occlusal aspect of the alveolar crest in order for occlusal bone formation to occur.

After an 11 year follow-up, Denissen *et al.* (1989) reported that during the first observation period of five years, 16 of 81 bulk HA implants placed in 11 patients had become exposed but were ankylosed to the bone and therefore could not be removed. Of these 16 implants, four eventually became loose and were lost (4.9%). In the second observation period (5 to 11 years) 16 implants of the remaining 77 were lost. This made a total of 20 implants lost, or 24.7% (Fig. 15.1). Ten implants still present became exposed although ankylosed. All implants had been placed into fresh empty sockets of the mandible, the gingivae sutured, and immediate dentures placed. The survivability results of this study, combining percentage of implants lost and percentage of implant exhibiting dehiscence, is plotted in Fig. 15.1 and compared in Figs. 15.2 and 15.3 with other clinical reports of HA cone implants.

Encouraged by the early reported success of HA cones in the Denissen *et al.* studies, Cranin and Shpuntoff (1984), working with 10 dental residents, placed 100 dense HA non-resorbable cones into fresh sockets of 10 patients without the use of a flap procedure. The cones were placed at least 1 mm below the alveolar crest. Sixty two cones were placed in the maxilla and 38 in mandible. In contrast to Denissen and de Groot's favorable results, at the end of one year 55% of the cones were lost: 32 (51.6%) maxillary cones and 23 (60.5%) mandibular cones (Fig. 15.1).

Results from numerous other investigators fall between these two extremes. Figure 15.1 summarizes the time dependence of the percentage of dense hydroxyapatite implants lost or exhibiting dehiscence for various reported trials. The reference citations of the studies are included in Fig. 15.1. Figs. 15.2 and 15.3 separate the clinical performance of HA cones as percentage loss (Fig. 15. 2) and percentage of dehiscence (gradual exposure of the implant) (Fig. 15.3). In many instances dehiscence can be repaired by grinding away the exposed surface of the implant. However, this is usually undesirable and can lead to progressive failure of the implant.

The results for HA implants shown in Fig. 15.2 show a low of zero loss after 12 months (Denissen and de Groot, 1979) and a high of 55% for the same time period (Cranin and Shpuntoff, 1984). Three other studies showed

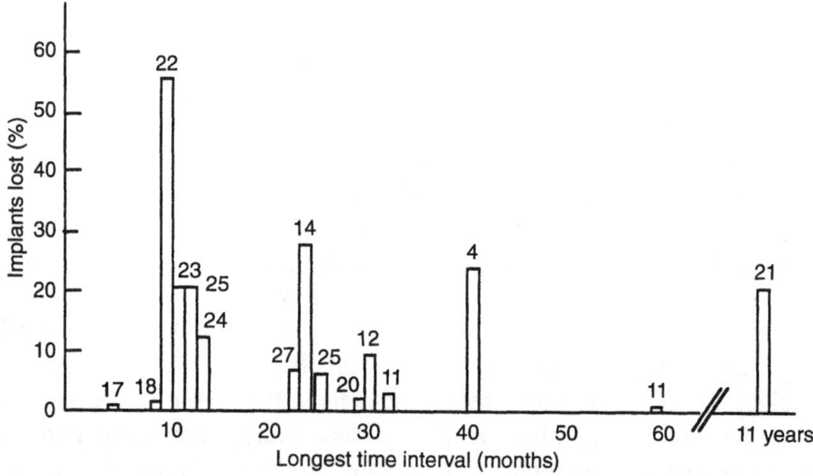

Fig. 15.2 Percentage of HA cones lost for endosseous alveolar ridge maintenance of denture wearers.

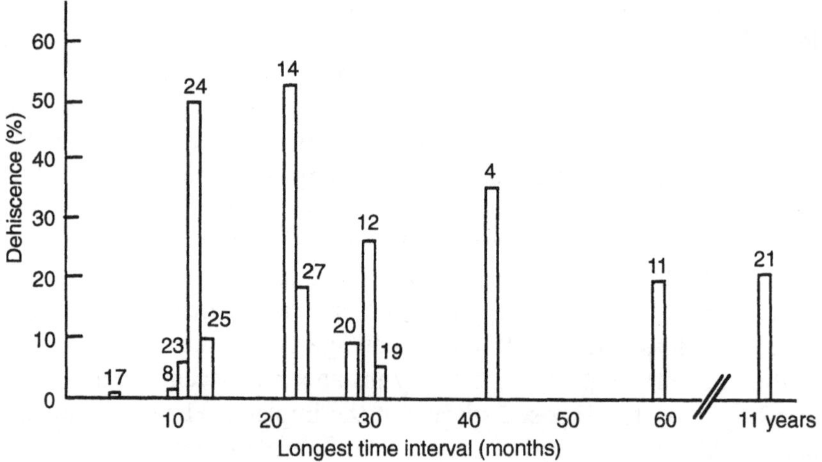

Fig. 15.3 Dehiscence of HA cones used for endosseous alveolar ridge maintenance of denture wearers.

results of 20% (Brook *et al.*, 1988), 20% (Sattayasanskul *et al.*, 1988), and 12% (Filler and Kentros, 1987) lost by 12 months. The percentage of dehiscence for the 12 months studies (Fig. 15.3) also show a low of zero percent (Denissen and de Groot, 1979) and a high of 50% (Filler and Kentros, 1987) with intermediate values of 9.5% (Sattayasanskul *et al.*, 1988) and 8.6% (Brook *et al.*, 1988). Two 24 months studies of HA cone implants showed 6.1% lost and 17.7% dehiscence (Kangvonkit *et al.*, 1986) and 27% lost and 53% dehiscence (Kwon *et al.*, 1986). Quinn *et al.'s* (1985) 31 months study of HA cone implants resulted in 9.7% lost and 26% dehiscence whereas Bell (1986) reported 23.5% lost and 35.3% dehiscence at 42 months. After 11 years Denissen *et al.'s* (1989) results were 24% lost and 30% dehiscence.

This wide range of failure or success of HA cone implants is due to many factors which vary from center to center. The variables affecting clinical performance include: fit, technique, shape of implant, contouring of the implant at time of surgery, primary closure, length of healing period prior to fitting of denture, condition of the alveolar bone at time of surgery, subsequent oral hygiene, etc.

The clinical studies of HA implants show that the concept of prevention of alveolar ridge resorption is valid for implants that remain in place. In the Quinn *et al.* (1985) study the patients had approximately twice as much alveolar bone maintained according to height and width measurements as compared to control sites. After 23 months, ridge height and width loss was 2.7 and 2.6 mm respectively in the implant group vs 5.5 and 4.5 mm in the control group. Kangvonkit *et al.'s* (1986) findings were similar; after two years there was only 1.4 mm ridge height resorption in the implant group vs 4.2 mm in the control group. Sattayasanskul *et al.* (1988) also report, at one year, significantly less ridge resorption (2.5 mm) for the implant group vs the control group (4.1 mm). However, the 21 month study of Kwon *et al.* (1986) did not yield a statistically significant difference between implant (3.3 ± 2.8 mm) and control (5.7 ± 0.8 mm) groups. During the follow-up period dentures were relined an average of five times for the implant group and three times for the control group.

The wide variation in clinical performance of HA implants, as illustrated in Figs. 15.1, 15.2 and 15.3, make use of HA for retention of the alveolar ridge of questionable clinical value.

BIOGLASS® CONE IMPLANTS

Bioglass® is another promising material used as cone implants to prevent alveolar ridge resorption. The composition of 45S5 Bioglass® is 45% SiO_2, 24.5% CaO, 24.5% Na_2O and 6% P_2O_5 (by weight). The material developed in 1969 (Hench *et al.*, 1971) has several unique characteristics:

1) It bonds rapidly to both cortical and cancellous bone (Hench and Clark, 1982).
2) It forms an adherent bond with soft connective tissues (Wilson and Noletti, 1990).
3) It develops an inorganic interfacial reaction zone of 200-300 μm thickness composed of a hydrated silica gel layer and a hydroxy carbonate apatite layer bonded to tissues. The inorganic bonding zone mimics the mechanical properties of the periodontal membrane (Wilson *et al.*, 1993; Weinstein *et al.*, 1980).
4) It stimulates the proliferation of bone due to activation of bone stem cells which leads to interfacial bonding within a few weeks of implantation (Hench, 1994).

Stanley *et al.*, (1986) reported on an 18 month clinical study using 45S5 Bioglass® cone implants, termed endosseous ridge maintenance implants (ERMIs) (Fig. 15.4). This clinical trial followed a 2 year study in baboons where it was discovered that the tissue bonding of Bioglass® tooth root implants resulted in a stable alveolar ridge (Stanley *et al.*, 1981). In the human trial at 18 months only 1.8% of the 216 cones placed in 26 patients had been lost and only 2.8% had developed dehiscence (Fig. 15.1).

A follow-up study of 32 months (longest time interval) of the 45S5 Bioglass® cone implants was published and the results remained excellent

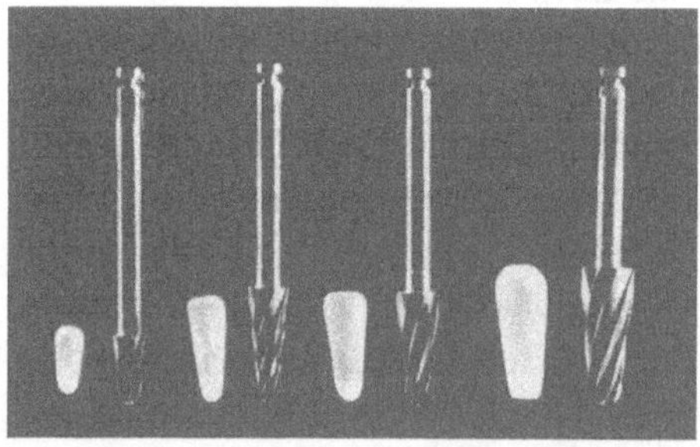

Fig. 15.4 Four Bioglass® endosseous ridge maintenance implants (ERMI) with matching dental burrs.

with only 2.9% implants lost and only 3.7% dehiscence (Stanley *et al.*, 1987). Based on these favorable clinical results the 45S5 Bioglass® cones became commercially available. This study has continued with 242 cone implants of 45S5 Bioglass® in 29 patients.

In 1994 Stanley *et al.*, published the results of a long-term follow-up of 20 of the original 29 patients consisting of 12 males and 8 females; with a mean age at the time of surgery of 42.3 years (range 24-76 years). Originally a total of 168 implants were placed in these patients; fourteen received complete maxillary/mandibular dentures and six received the combination of a complete maxillary denture and a mandibular removal partial denture. After an average post-implant period of 54.13 months (range 36-71 months), a total of 21 (12.5%) implants had been lost (average 30.9 months) and 13 (7.7%) had been recontoured (average 17.0 months) (range 4-64 months). Of 83 cones placed in the maxilla, 3 (5.3%) of 57 in anterior sites and 3 (17.6%) of 17 in posterior sites were lost. Of 85 placed in the mandible, 9 (17.3%) of 52 were lost from anterior sites and 4 (12.1%) of 33 from posterior sites.

Generally, the anterior mandible has the highest implant survival rate and the posterior maxilla the lowest survival rate, success usually declining when implants are placed in more porous bone. In the 20 recalled patients the maxillary anterior implants had the highest implant survival rate followed by the mandibular posterior implants, the maxillary posterior implants, and the mandibular anterior implants.

Recontoured implants

Thirteen (7.7%) of 168 implants were recontoured because of dehiscence between 4 and 46 months (mean 17.0 months) after initial implantation. Two implants were contoured at 4 months and again at 39 months. In the maxilla, three of 82 implants (3.7%) required recontouring; 2 of 56 (3.6%) in anterior sites and one of 26 (3.8%) in a posterior site. None of the recontoured implants in the maxilla was lost.

In the mandible, ten of 86 implants (11.6%) required recontouring: two of 52 (3.8%) in anterior sites and eight of 34 (23.5%) in posterior sockets. One recontoured mandibular posterior implant was recontoured at seven months and lost at 14 months.

Relining dentures

The need for a reline was not unexpected since the Bioglass® cones were placed approximately (and intentionally) 2.0 to 3.0 mm below the superficial alveolar bone which might then be resorbed. Seventeen of the 20 recalled patients had need of a reline. The time interval between insertion of dentures and the decision to reline varied from 4 months to 23 months (average 14.5). The average time interval for a reline of the maxillary complete denture in patients with a removable mandibular partial denture was shorter (10.9 months) than the average interval for a reline in patients with two complete dentures (14.4 months).

Early use of dentures to slow down the rate of ridge resorption after cone implantation does not appear to be important. The initial early rate of alveolar ridge resorption which takes the remaining bone down to the level of the implanted cones continues, regardless of whether the dentures are placed sooner or later.

The results of Fig. 15.1 show that Bioglass® cones will probably succeed in any location provided enough bone is present and they are placed deeply enough.

MATERIALS DIFFERENCES

Differences in clinical methods can greatly affect clinical success in alveolar cone implants. The results of Stanley *et al.*, (1987, 1994) have been excellent over an extended post-implant period in part because Bioglass® cone implants were placed deep and dentures were not inserted until after 6 weeks to 4 months. Special dental burs were also developed that matched the shape and dimension of the cone implants (Fig. 15.4). Use of the burs at low speed to ream minimally the tooth socket prior to insertion of the implant ensured a close fit and may be an important contribution to the high survivability. The lower resorption rate in the maxilla may explain part of the success of Bioglass® implants compared to those studies that used only mandibles and immediate dentures.

It appears that a Bioglass® cone implant with a good, even without an exact, surgical fit in bone may initiate enough partial or patchy direct contact between the implant and cortical bone that it will suffice to stabilize the implant. The more exact the surgical fit the more appositional lamellar bone will form along the entire implant surface and improve fixation. Because of the gel interfacial bonding layer, there is no need for microscopic or gross surface configurations–the smoother the surface, the better the result. Nevertheless, there are also important material differences between HA and Bioglass® implants that can also affect their behavior. 45S5 Bioglass® implants develop a bond to both soft and hard tissues whereas HA bonds only to bone. The differences in soft and hard tissue bonded 45S5 Bioglass® cones buried in modified extraction sites in dogs has been studied (Wilson *et al.*, 1993). The methods, implants and dental burs used in the canine model were the same as those procedures used in the human trials. Within 3 months the bonding had stabilized for both hard and soft tissues. The soft tissue was bonded by collagen fibers interdigitated in a 150-400 μm thick bonding gel layer composed of biological hydroxycarbonate apatite and an underlying silica-rich gel layer that began to form on the cones within minutes of implantation (Fig. 15.5). This thickness was nearly 50% greater

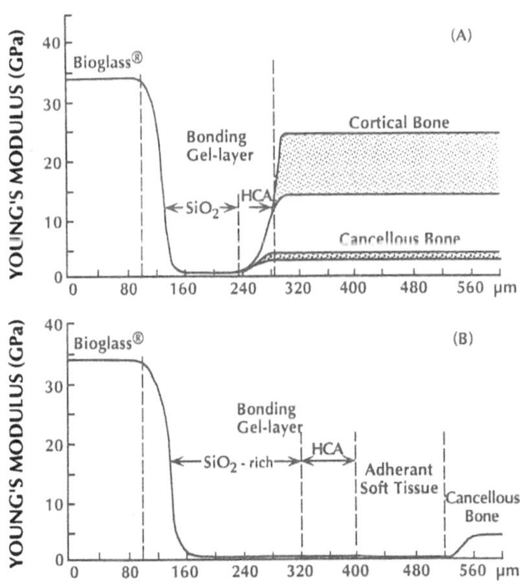

Fig. 15.5A Elastic modulus gradient of Bioglass® ridge maintenance implant and cortical bone. From Wilson *et al.*, 1993 with permission.

Fig. 15.5B Elastic modulus gradient of Bioglass® ridge maintenance implant and soft tissue. From Wilson *et al.*, 1993 with permission.

than the bonding zone between bone and the implant. The thickness of the bonding zone to either bone or soft tissue was stable after about 7 months (Fig. 15.6).

The elastic compliance of this bonding zone and the favorable stress transfer resulting from the bonding of collagen fibrils (Wilson and Noletti, 1990; Weinstein *et al.*, 1980) may well be related to the short and long term term success of cone implants of 45S5 Bioglass®. In contrast, HA implants bond much more slowly and form a bonding zone only to bone with a thickness of less than 1 μm, resulting in a much less natural stress transfer gradient into the alveolar ridge (Fig. 15.7). These biomaterials-biomechanics factors may contribute to the early 25 to 50% failure rates of HA cones. The conclusion seems to be that HA is much more sensitive to surgical technique and fit because the biomaterial has a much lower index of bioactivity, and its retention cannot be enhanced by soft tissue adhesion.

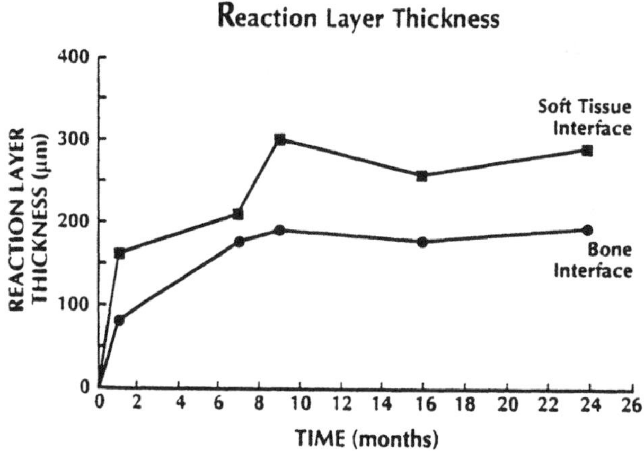

Fig. 15.6 Development of the reaction layers on Bioglass® ridge maintenance implants. From Wilson *et al.*, 1993 with permission.

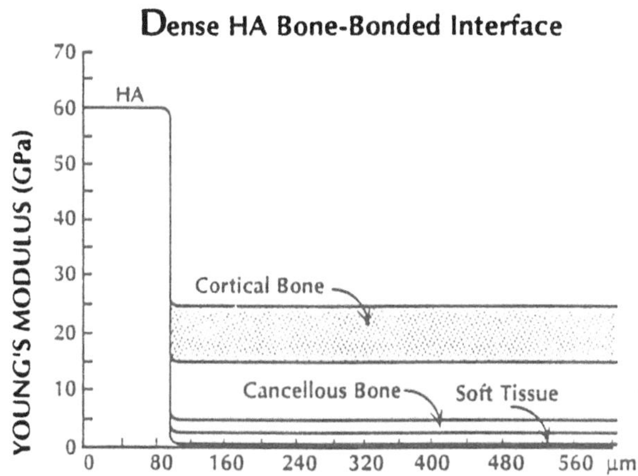

Fig. 15.7 Elastic modulus gradient of dense HA and bone. From Wilson *et al.*, 1993 with permission.

REFERENCES

Atwood, D.A. (1962) *J. Prosth. Dent.* **12**, 441-50.

Atwood, D.A. (1971) *J. Prosth. Dent.* **26**, 266.

Atwood, D.A. (1979) *J. Periodont.* **50**, 11-21.

Bell, D.H. (1986) *J. Prosth. Dent.* 323-26. **(4)**

Brook, I.M., Sattayasanskul, W. and Lamb, D.J. (1988) *Br. Dent. J.* **164**, 212. **(23)**

Cranin, A.N. and Shpuntoff, R. (1984) *J. Dent. Res.* **63** (269), 200. **(22)**

Denissen, H.W. and de Groot, K. (1979) *J. Prosth. Dent.* **42**, 551-56. **(18)**

Denissen, H.W., Kack, W., Veldhuis, A.A.H. and van den Hooff, A. (1989) *J. Prosth. Dent.* **61**, 406-712. **(21)**

Denissen, H.W., Rjeda, B.V. and de Groot, K. (1978) *Trans. SFB* **II**, 188. **(17)**

Denissen, H.W., Veldhuis, A.A.H., Makkes, P.C., *et al.* (1980) *Clin. Prev. Dent.* **2**, 23-28. **(20)**

Filler, S.J. and Kentros, G.A. (1987) *J. Dent. Res.* **66**, 249. **(24)**

Hench, L.L. (1994) Bioactive ceramics: Theory and applications, in *Bioceramics 7*, (eds Ö.H. Andersson and A. Yli-Urpo), Butterworth-Heinemann Ltd., Oxford, England, pp. 3-14.

Hench, L.L. and Clark, A.E. (1982) Adhesion to bone, in *Biocompatibility of Orthopaedic Implants*, Vol II (eds D.F. Williams and G.D. Winter), CRC Press, Boca Raton, Florida, Chapter 6.

Hench, L.L., Splinter, R.J., Greenlee, T.K. and Allen, W.C. (1971) Bonding mechanisms at the interface of ceramic prosthetic materials, *J. Biomed. Mater. Res.* **2** (1), 117-41.

Hench, L.L., Stanley, H.R., Clark, A.E., *et al.* (1991) in *Bioceramics 4,* (eds W. Bonfield, G.W. Hastings and K.E. Tanner), Butterworth-Heinemann Ltd., Oxford, England, pp. 231-38.

Kangvonkit, P., Matukas, V.J. and Castleberry, D.J. (1986) *Int. J. Oral Maxillofac. Surg.* **15**, 62-71. **(27)**

Kwon, H.J., El Deeb, M., Morstad, T. and Waite, D. (1986) *J. Oral Maxillofac. Surg.* **44**, 503-8. **(14)**

Lam, R.V. (1972) *J. Prosth. Dent.* **27**, 311-23.

Misch, C.E. (1994) *J. Mich. Dental Assoc.* **66**, 219-23.

Quinn, J.H. and Kent, J.N. (1984) *Oral Surg.* **85**, 511-21.

Quinn, J.H., Kent, J.N., Hunter, R.G. and Schaffer, C.M. (1985) *J.A.D.P.* **110**, 189-93. **(12)**

Rothstein, S.S. (1984) *J. Am. Dent. Assoc.* **109**, 571-74.

Sattayasanskul, W., Brook, I.M. and Lamb D.J. (1988) *Int. J. Oral Maxillofac. Implants* **3**, 203-7. **(25)**

Smiler, D.G. (190) *Implant Digest* **Summer,** 2-3.

Sobolik, D.F. (1980) *J. Prosth. Dent.* **50**, 612-19.

Stanley, H.R. (1987) in *The Dental Annual*, John Wright & Sons, Bristol, pp. 209-30.

Stanley, H.R. *et al.* (1995) Alveolar ridge maintenance using endosseously placed Bioglass® 45S5 cones, in *Encyclopedia of Biomedical Materials*, Marcel Dekker Inc., New York, pp. 2707-63.

Stanley, H.R., Clark, A.E., Hall, M.B., *et al.* (1986) *Trans. Soc. for Biomaterials* **IX**, 150.

Stanley, H.R., Hall, M.B., Colaizzi, F. and Clark, A.E., (1987) *J. Prosth. Dent.* **58**, 607-13.

Stanley, H.R., Hench, L.L., Bennett, C.G. *et al.* (1981) *Int. J. Oral Implant* **2**, 26-36.

Stanley, H.R., Brandt, R.L., Hall, M.B., Clark, A.E. and King, C. (1994) *J. Dent. Res.* **73** (special issue) 399, Abstract #2376.

Tallgren A. (1967) *Acta Odontol. Scand.* **25**, 563-592.

Tallgren A. (1969) *Acta Odontol. Scand.* **27**, 539-61.

Veldhuis, H., Driessen T., Denissen, H. and de Groot, K. (1984) *Clin. Prev. Dent.* **6**, 5-8. **(11)**

Weinstein, A.M., Klawitter, J.J. and Cook, S.D. (1980) *J. Biomed. Mater. Res.* **14**, 23-29.

Wilson, J., Clark, A.E., Hall, M. and Hench, L.L. (1993) *J. Oral Implantology* **XIX** (4), 295-302.

Wilson, J. and Nolletti, D. (1990) in *CRC Handbook of Bioactive Ceramics,* CRC Press, Baca Raton, FL, pp. 1.

Wilson, J., Pigott, G.H., Schoen, F.J. and Hench, L.L. (1981) *J. Biomed. Mater. Res.* **15**, 805-17.

16

Summary

Larry L. Hench
June Wilson

GENERAL CONCLUSIONS

Several conclusions can be made from the analyses of clinical performance of the various skeletal prostheses discussed in Chapters 2-15.

Evaluation of all categories of prostheses shows that standardized methods of reporting clinical performance of devices is needed. Many hospitals use their own performance scales thereby making it difficult to compare results even with the same prosthetic systems. The effort to coordinate clinical performance scales in this volume is a poor substitute for having common evaluation procedures as a basis for comparison.

Often clinical studies do not report age, gender, initial condition of patient, and other factors that can affect clinical performance. This lack of data makes it difficult to assess the basis for the differences in performance of the prosthetic systems.

For many types of prostheses clinical success is high for the first five years. Thus, it is especially important that long term (>10 years) data be reported. All too frequently general clinical use of new types of devices occurs before long term comparative data is available to establish statistically significant improvement over devices with long term performance data.

TOTAL HIP PROSTHESES

The cemented low friction (Charnley-type) total hip arthroplasty (THA) using a metallic femoral component and UHMWPE cup has the highest level of clinical success, as illustrated in Fig. 16. 1. The Kaplan-Meier survivability curve shown in Fig. 16.1 (Chapter 2) is based upon 42 published clinical

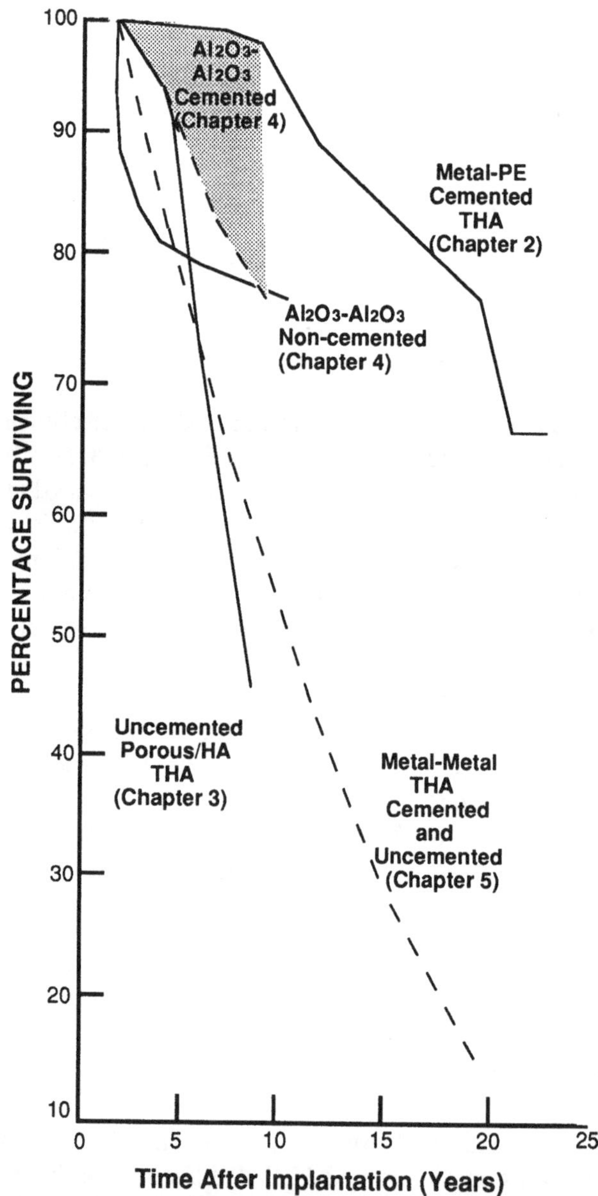

Fig. 16.1 Survivability curves for total hip arthroplasties.

studies involving a total of 15,051 patients for time periods as long as 25 years.

The predicted survival rates for the cemented low friction type of THA are:

5 years 99.41 ± 0.02%
10 years 95.48 ± 0.04%
15 years 83.12 ± 0.18%
20 years 66.53 ± 0.35%

A lifetime of 19 years is expected for the average arthroplasty using the cemented low friction prosthesis.

It is uncertain whether the fall off in survivability between 15 and 20 years will continue during the next five years. This is because of the relatively small number of long term studies reported and the significant improvement in clinical techniques developed during the last fifteen years. Advances in surgical procedures should lead to improved 15-25 year lifetimes for cemented THAs.

Life expectancy of THAs depends upon the following factors:

- gender (males lower than females);
- weight (overweight lower than normal);
- age (younger patients with high activity lower than older).

Thus, low friction THA has a low risk of failure for patients >65 years of age and should be considered to be the standard for hip replacement.

Other types of THA systems generally show lower survival rates than the cemented low friction type, as shown in Fig. 16.1. The alumina-alumina cemented prosthesis is the exception. However, the distribution of results and relatively low number of clinical studies reported for alumina-alumina prostheses (Chapter 4) makes it impossible to establish statistical evidence that this system improves THA survivability by eliminating UHMWPE wear debris.

Uncemented THA systems, regardless of means of fixation, show lower survivability results (Chapters 3 and 5). The difference appears by five years and expands with time (Fig. 16.1). The difference between cemented and uncemented THA shown in Fig. 16.1 is confirmed by the Havelin *et al.*

study of 14,000 patients which shows two times the number of failures of uncemented hips by 5 years.

TOTAL KNEE PROSTHESES

All total knee arthroplasty (TKA) systems exhibit generally similar survivability results of 82-90% at 10 years (Chapter 6).

There is no statistical difference between cemented and uncemented TKAs.

SHOULDER PROSTHESES

There is no statistically significant difference between cemented and uncemented shoulder prosthetic systems (Chapter 7).

Prior condition of the patient and associated soft tissue defects are the primary factors for success of a shoulder prosthesis.

If a shoulder prosthesis is essential, the results show that an unconstrained hemi-arthroplasty or unconstrained total arthroplasty is preferred.

ELBOW PROSTHESES

All types of elbow prostheses provide marked improvement in quality of life, less pain, and more range of motion (Chapter 8).

Failures of elbow prostheses are uncommon (10-15%) for most types of implants but complications range from 30 to 80%. There are insufficient data to establish statistically significant differences between the types of prostheses used.

TOE PROSTHESES

Most prostheses are successful with little difference in clinical performance (Chapter 9).

LIMB LENGTHENING

All external systems used for limb lengthening have complications (Chapter 10).

New systems using intramedullary devices that minimize infection and psychological complications show promise of improved success but require additional time of evaluation.

SPINE

There are no common rating systems between plated and non-plated methods, this makes it difficult to compare relative success (Chapter 11).

There are generally equivalent results of approximately 90% success for plated vs no-plate methods.

VENTILATION TUBES

No current material or device provides long term (>2 years) aeration of the middle ear without damage to the tympanic membrane (Chapter 12)

Short term (<2 years) results depend upon the design of the ventilation tube.

OSSICULAR REPLACEMENT PROSTHESES

Bioactive implants provide longer term success than bioinert implants (Chapter 13).

OSSEOINTEGRATED DENTAL IMPLANTS

There are no significant differences in the success rates among the various implant systems reported in 25 clinical studies (Chapter 14).

No study conformed to the NIH criteria for dental implants.

The criteria for success and failure differed so much between studies that statistical comparisons were not possible.

Because of variability in criteria for success, the reported results of 82-100% success for the various reports must be viewed with caution.

ALVEOLAR RIDGE MAINTENANCE IMPLANTS

Bioactive glass implants inserted surgically into the alveolar ridge following reaming with a specially designed dental burr provided much lower rates of dehiscence and fewer implants lost over 5-10 years than synthetic hydroxyapatite implants (Chapter 15).

FINAL CONCERN

A concern felt by the authors of almost all the chapters was for the objectivity of investigators responsible for the development and testing of particular devices. This was often reflected in the experimental design in which the groups were unbalanced and which appeared to produce a bias toward the device or system being evaluated. However, it is to be expected that the originator of a system would be expected to have a rather more meticulous approach to its application than later users. It is important to consider this factor when evaluating the relative performance of various prosthetic systems.

Index

Zeitfracht Medien GmbH
Ferdinand-Jühlke-Straße 7
99095 Erfurt, Deutschland
produktsicherheit@kolibri360.de